GW01271838

GUATEMALAN BACKSTRAP WEAVING

UNIVERSITY OF OKLAHOMA PRESS : NORMAN

GUATEMALAN BACKSTRAP WEAVING

BY

NORBERT SPERLICH

AND

ELIZABETH KATZ SPERLICH

WITH PHOTOGRAPHS AND DRAWINGS BY THE AUTHORS

Library of Congress Cataloging in Publication Data

Sperlich, Norbert, 1938–
Guatemalan backstrap weaving.

Bibliography: p. 171
Includes index.
1. Indians of Central America—Guatemala—Textile industry and fabrics. 2. Hand weaving—Guatemala. I. Sperlich, Elizabeth Katz, 1944– joint author. II. Title. III. Title: Backstrap weaving.
F1465.3.T4S63 677'.028242'097281 79–26991

Manufactured in the U.S.A.
First edition.

»»»»»««««««

CONTENTS

»»»»»»»»»»»»»»»»»»»»»«««««««««««««««««««««

ILLUSTRATIONS

COLOR PLATES

BLACK-AND-WHITE ILLUSTRATIONS

MAP

»»»»»«««««

PREFACE

Our observations on backstrap weaving were made during a stay of eight months (from January to September, 1976) in the highlands of Guatemala. Living with Indian families, we learned the basics of backstrap weaving, studied weaving techniques, and made a collection of textiles. We visited more than half of the forty or so towns where backstrap weaving is practiced. Since the different costumes of Guatemala have been described by several authors, we decided to concentrate our efforts on technical aspects of backstrap weaving. We describe most of the basic techniques used by backstrap weavers in Guatemala.

Norbert Sperlich
Elizabeth Katz Sperlich

Grandmother and grandchildren from Chajul.

»»»»»»»»»»»»»««««««««««««

INTRODUCTION

About half of the people of Guatemala are Indians. Their cultural heritage—language, beliefs, and way of life—sets them apart from the Ladinos or non-Indians. Another factor that separates most Indians from Ladinos is the lower social status of the Indians. From the days of the conquest until recent times the Indians have been kept in submission and servitude. Until 1945 there existed discriminatory laws obliging Indians to work periodically on the plantations of rich landowners. Even today the Indians have not yet managed to free themselves from their role as second-class citizens.

THE INDIANS OF GUATEMALA

Many Indians live in rural communities of the highlands and identify strongly with the local culture of their *municipio*. A *municipio* is formed by a town *(pueblo)* and a number of smaller settlements *(aldeas)*. Indians from one *municipio* regard themselves as a group set apart from other *municipios*. They often differ in costume, customs, and dialect or language from their neighbors.

Most *municipios* originated in the sixteenth century when the Indians were forced by the Spaniards to settle in towns. Each town was given a sufficient amount of land that was owned communally but cultivated individually. The local government was in the hands of the Indians, supervised by Spanish officials and priests. Periodically people had to leave their hometowns to work on the plantations, in mines, and in the cities of the Spaniards. For their own upkeep they cultivated the communal land and engaged in a number of crafts. A part of their production was used to pay tribute to the Spanish king.

Through the centuries each town developed its local version of a culture that blended native and European elements. Worship of the patron saint and participation in a complex system of ranked public offices strengthened the bonds between the Indians and their hometowns.

Toward the end of the eighteenth century the power of the Spaniards declined. During the following period of political and economical instability less control was exercised over the Indians than before. When Central America declared its independence in 1821, the status of the Indian communities did not change much. In 1871 the liberals came into power in Guatemala. An agrarian reform, aimed at increasing the number of landowners and promoting cash-crop agriculture, was introduced in favor of middle-class Ladinos. (In colonial times the term Ladino was employed for Spanish-speaking persons of mixed blood. Ladinos had more privileges than the Indians, but fewer than the descendants of the Spaniards.) Most communal land of the Indian towns was transformed into private property. Some of this land was claimed by Ladinos who settled in Indian towns. Ladinos also took over important posts in the local government of the Indian *municipios*. Coffee became the most important commodity for export, and growing numbers of workers were needed on the plantations. A new era of oppression began for the Indians. As in colonial times, they were forced to work for little pay. Forced labor on the plantations was abolished after the revolution of 1944. Under President Jacobo Arbenz farm workers started to organize, and an agrarian reform was under way. These promising developments came to an end when the Arbenz government was overthrown in 1954.

Traditionally farming has been the mainstay of the Indians, but today overpopulation of the highlands, lack of land, and primitive methods of agriculture make it impossible for most Indians to "live off the land." On the other hand, huge estates in the fertile coastal areas are concentrated in the hands of a small number of rich landowners. Several hundred thousand Indians are employed there on the cotton and coffee plantations during harvest time. Many landless Indians live permanently on the plantations, and more and more people depend on seasonal work. Working conditions on the plantations are bad, and the pay is low. Other possibilities of employment are few. Most of the Indians are illiterate, and even the small number who have an education find it difficult to get work because there are not enough jobs available. Fearful of growing social unrest, the national governments of Guatemala have in recent times made some half-hearted attempts to improve the plight of the Indians. Farmers and craftspeople are encouraged to organize in cooperatives. However, where the interests of the big landowners are at stake, the government does not hesitate to send troops against the Indian peasants.

The traditional Indian culture, centered around farming and the social and religious life of the community, is undergoing many changes. People are willing to try new methods of farming, and they will accept innovations that are advantageous to them. Many Indians are organized in agricultural or savings and loan cooperatives. More and more children attend school, even though their work is needed in the household or in the fields. Most Indian men have given up their traditional costumes, but women are still expected to wear the local variation of the typical Indian dress. Of the traditional crafts backstrap weaving is still flourishing in many *municipios*. Pottery is made in a few places.

Whereas in the past Indians of one *municipio* were united in their religious beliefs, we now find them divided into factions. The traditionalists who worship saints as well as pagan deities are under attack from Roman Catholic missionaries and their followers. Various Protestant churches have also gained a number of adherents.

Economical differences among the Indians are more pronounced than in the past. People who run a store or contract workers for the plantations are often better off than the mass of the farmers. However, most of the storekeepers and contractors are Ladinos. There exist strong resentments between Indians and Ladinos. Some Indians despise everything Ladino, others are willing to accept some of the trappings of Ladino culture, and a few consider their Indian heritage a burden.

In their *municipios* Indians have regained some of the ground that was occupied by Ladinos. An Indian mayor (instead of a Ladino mayor) in a predominantly Indian *municipio* is no longer an exception. On the national level, however, there are no political parties or other organizations that represent the interests of the Indians.

Mother and daughter from Nebaj.

INDIAN COSTUMES

Most of the Indian women, and in some *municipios* the men too, dress in the so-called *traje típica*, a costume that identifies them as belonging to the Indian population of a certain town. Where the local costume has disappeared, women buy skirts and blouses that are definitely Indian in style, but not necessarily typical for a certain *municipio*.

In the past women wove clothing for themselves and their families on the native backstrap loom. Today some parts of the *traje típica* are made from material woven on the foot loom or in a factory.

Basically, a woman's costume consists of a huipil (a kind of blouse), a skirt, and a belt or sash. In addition many women wear some kind of headdress and a rectangular piece of cloth *(perraje)* to use for carrying bundles or as protection against the cold. In their basic form these parts of the woman's costume were already in use before the arrival of the Spaniards. Huipils are still woven on the backstrap loom in about fifty *municipios*. Weavers in the Department of Huehuetenango also weave their own skirts; elsewhere, skirts are made from material woven on the foot loom.

The traditional men's costume, much influenced by European models, consists of pants, shirt, sash, *tzute* (a rectangular piece of cloth for covering the head), a wool coat or jacket, and—in some places—a wool blanket that is wrapped around the hips. Large *tzutes* are also employed to carry bundles, especially in the Department of Huehuetenango. As with the women's costume, shirts and pants are made from rectangular panels and have a very simple cut. Men's clothing of the traditional kind is woven on the backstrap loom except for woolen articles, which are woven in most places on treadle looms.

Today most Indian men wear "modern" clothing. The traditional costumes still persist, however, in the Lake Atitlán area, in the *municipios* of Nahualá, Santa Catarina Ixtahuacán, San Martín Sacatepéquez, San Juan Atitán, Todos Santos and in some remote *aldeas* of other *municipios*. When switching to the Ladino (non-Indian) way of dressing, men in rural areas often retain some parts of their old costume. The traditional sash and, in some areas, the *capixay*, a woolen coat, are often worn together with factory-made garments. Children are dressed much like their parents.

Costumes can differ greatly from one *municipio* to the next. When and how the many different costumes of Guatemala originated is not known. Chroniclers in colonial times did not speak of Indians from different villages dressing in different costumes. Instead we get the impression from these writers that the mass of the Indian people wore very humble clothing that differed little in color and style. Upper-class Indians, who conducted the government of a town, wore more elaborate clothing, which was often modeled after the dress of the Spaniards. It is possible that after the colonial period came to an end, some common people started to copy designs and symbols from the costumes of the upper class, creating a style of dress that in time was accepted by all the inhabitants of a *municipio*.

Toward the end of the last century large numbers of Indians were forced to work on the newly established coffee plantations, and much Indian land was taken over by Ladinos. In many areas the traditional life style of the Indians was disrupted, and backstrap weaving declined or disappeared altogether. The use of material woven on the foot loom was increased, and industrial yarn became available to backstrap weavers. Today weavers can choose from many kinds and colors of industrial yarn, including acrylic yarn with Day-Glo colors, and Lurex threads. As traveling merchants bring yarn of all colors to even the most remote villages, textiles tend to become more gaudy. Weavers experiment with new patterns or copy patterns from other *municipios*.

BACKSTRAP WEAVING

With the spread of factory-made articles, local crafts and industries have been declining in Guatemala. However, weaving is still an important craft in large areas of the highlands. The treadle loom, introduced by the Spaniards, is used mainly by

Girls from Nebaj.

men who are professional weavers. Women weave on the much simpler backstrap loom, an apparatus that the Indians have known since remote times. Even though weaving on the treadle loom is faster, the backstrap loom has many features that make it more suitable to the needs of an Indian household. No special installation is required for the backstrap loom, and loom parts are inexpensive or can be made from local materials. The loom is portable and can be hitched to any tree or house post, in the shade or in the sun, whichever the weaver prefers. While doing her weaving, a woman can keep an eye on her children or chat with a neighbor. Each weaver in a family has her own loom sticks, and one often sees several weavers working side by side.

Backstrap weaving is considered women's work. Nevertheless, men are familiar with weaving procedures, and in some areas male backstrap weavers specialize in weaving wool belts. Girls learn how to weave at an early age. They see their mothers and older sisters at the loom, and soon they start to play weaving with little sticks and leftover threads. At age seven or eight some girls weave their first piece—a band or a little *servilleta* (napkin)—under the supervision of their mothers (see plates 37, 38, 43). By the time they are twelve or thirteen years old, many girls are skilled weavers.

Most women spend a few hours a day at the loom, whenever they can get away from other household tasks. In some places women still weave a large part of the family's clothing. In most *municipios*, however, huipils are the main article woven on the backstrap loom.

In the past women spun white or brown native cotton and wool for their weaving materials. Coloring was done with natural dyestuffs. Today homespun cotton has, with few exceptions, been replaced by commercial yarn.

Since factory-made clothing and material woven on treadle looms are available everywhere, one may ask why women still continue to weave? Both cultural and economic factors play a role. In many towns *costumbre*, or local custom, requires that women dress in the traditional way. Commercial huipils that are different in style from the local ones are not acceptable. A huipil woven on the backstrap loom according to local standards is not only more beautiful than commercial huipils but also less expensive and more durable. A homemade huipil will last for at least two years of hard wear. Most women can afford to own only two huipils at a time.

While most weavers make clothing for their own family, there are a number of women in each *municipio* who weave for other people. Often they are commissioned to weave material from yarn that has already been warped by the client. In San Sebastián Huehuetenango women are paid about two quetzales for weaving a huipil. A weaver from Colotenango told us that she gets one quetzal for weaving a skirt. In both instances the weavers earn at the most five centavos per hour. One quetzal is worth one United States dollar. The Indians who live in the Huehuetenango area are among the poorest people in Guatemala. It is possible that in other parts of the country weavers are paid a little better for their work.

In some towns weavers offer their products on the market or leave them with a merchant who has a sales stand. Especially in Chichicastenango and Chiché, many huipils are offered for sale on the market. Every year more tourists visit Guatemala. They are eager to buy Indian textiles at bargain prices. Some "gringos" buy directly from the weavers, but most tourists get their souvenirs from merchants who paid very little to the weavers. Selling textiles is certainly better business than making them. Weavers who want better prices for their work have started to organize in cooperatives. This enables them to buy yarn in quantity and to sell textiles at fixed prices.

Geographical Distribution of Backstrap Weaving

In the past backstrap weaving was probably done in most of the two hundred or so *municipios* that have a large Indian population. Today we find about forty *municipios* in the central and western highlands where a great many Indian women still weave. In many more places weaving is done by some of the women. Whether or not backstrap

Woman from San Juan Atitán.

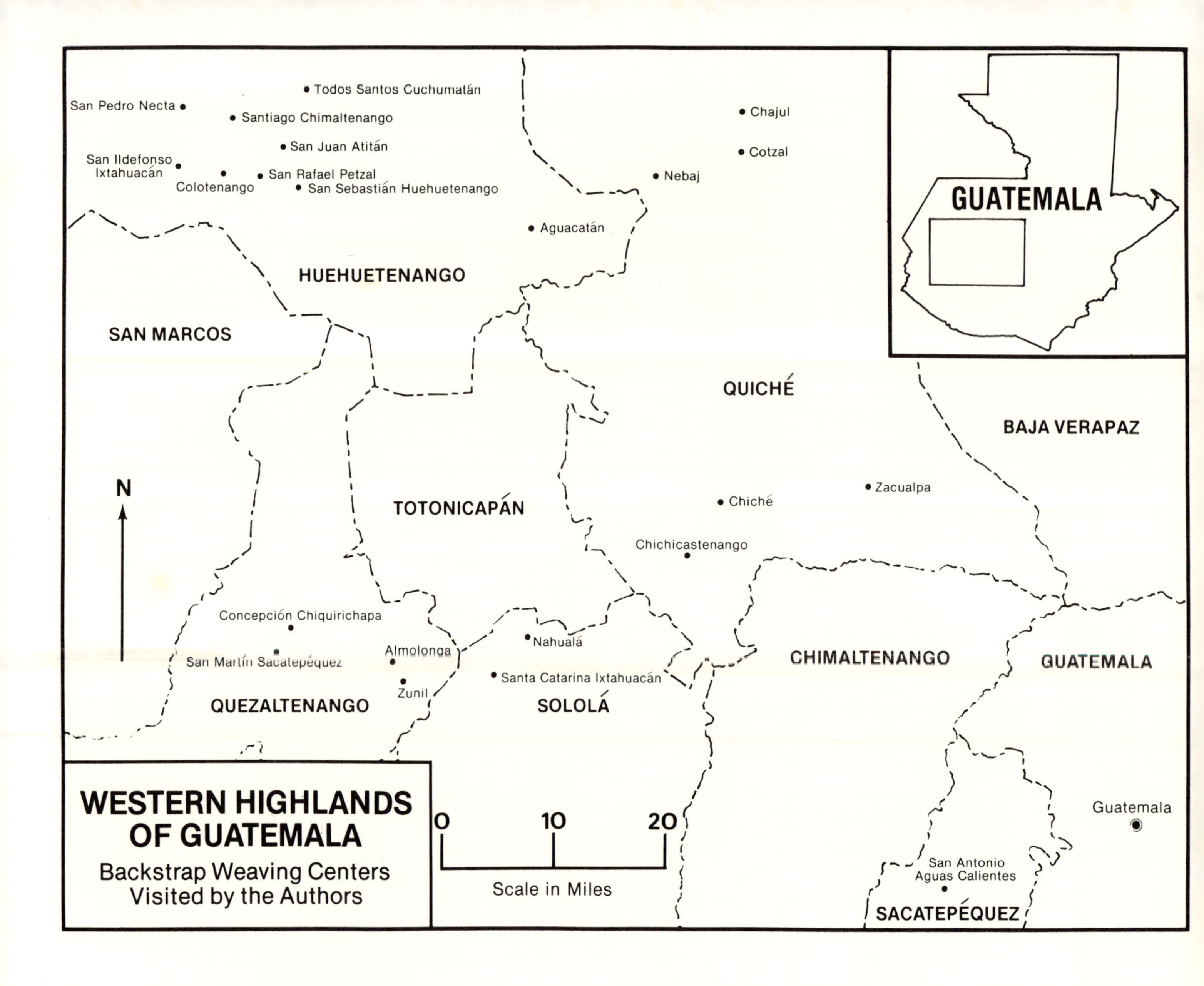

weaving is maintained in a community depends on many factors. Economic considerations play an important role. Weaving persists in areas close to the cities of Guatemala and Quezaltenango, whereas in some remote regions of the Cuchumatanes mountains it has long been abandoned. In the following chart we list all the *municipios* we visited, together with the most important garments that are woven there on the backstrap loom. Such items as blankets, *servilletas*, and bag straps do not appear on the list, even though they are made by weavers in several *municipios*. The map above shows the locations of the various towns.

Department of Huehuetenango

Only in the Huehuetenango area do women still weave skirts on the backstrap loom. In all the *municipios* listed below a great many of the women weave, perhaps with the exception of Aguacatán. In San Juan Atitán all garments for men, women, and children are locally made, except for women's headbands.

Municipio	Women's Clothing Woven Locally on the Backstrap Loom				
Aguacatán	......	Skirt	Belt	Headband	*Perraje*
Colotenango	Huipil	Skirt	Sash	....	*Perraje*
San Ildefonso Ixtahuacán	Huipil	Skirt	Sash		
San Juan Atitán	Huipil	Skirt	Belt	....	*Perraje*
San Pedro Necta	Huipil	Skirt	Sash	....	*Perraje*
San Rafael Petzal*	Huipil	Skirt	Sash	....	*Perraje*
San Sebastián Huehuetenango	Huipil	Skirt	Belt	....	*Perraje*
Santiago Chimaltenango	Huipil	Skirt	Belt	....	*Perraje*
Todos Santos†	Huipil				

*Costumes in San Rafael Petzal are similar to the ones in Colotenango.
†It is possible that some Todos Santos weavers make belts. Most women buy their belts from San Sebastián weavers.

Municipio	Men's Clothing Woven Locally on the Backstrap Loom				
Aguacatán	.....	.....	Sash		
Colotenango	.....	.....	Sash	*Tzute*	
San Ildefonso Ixtahuacán	.....	.....	Sash	*Tzute*	
San Juan Atitán	Shirt	Pants	Sash	*Tzute*	
San Pedro Necta	.....	.....	Sash	*Tzute*	
San Sebastián Huehuetenango	Shirt	Pants	Sash	*Tzute*	*Delantera*
Santiago Chimaltenango	.....	.....	Sash	*Tzute*	
Todos Santos	Shirt	Pants	Sash		

A *delantera* is a small wool blanket worn around the hips (see plate 13). Wool yarn is woven into *delanteras*, shirts, and pants in a few remote *aldeas* of San Sebastián Huehuetenango. Wool blankets are also woven on the backstrap loom.

Department of Quezaltenango

Municipio	Women's Clothing Woven Locally on the Backstrap Loom		
Almolonga	Huipil		
Concepción Chiquirichapa	Huipil	Headband	*Perraje*
San Martín Sacatepéquez	Huipil	....	*Perraje*
Zunil	Huipil	....	*Perraje*

Men's Clothing
San Martín Sacatepéquez: shirt, pants, and sash.

San Martín Sacatepéquez is the only *municipio* in the department where most men still wear a distinctive, handwoven costume. Some men in Zunil use as underwear handwoven pants of archaic cut. Sashes are also worn by some men in Zunil. In Concepción Chiquirichapa many women are professional weavers. They weave material for huipils, shirts, shawls, etc., which they sell to people from other towns or to local women who have given up weaving. In Almolonga and Zunil women have little time for weaving, because they work in their vegetable gardens. Local semiprofessional backstrap weavers provide their neighbors with material for huipils and *perrajes*. Women in Almolonga, Concepción, and San Martín wear wool belts that are made by specialists in Ostuncalco. Belts typical of the colorful Zunil costume come from Cantel.

Department of Quiché

Municipio	Women's Clothing Woven Locally on the Backstrap Loom			
Chajul	Huipil	Sash	Headband	*Perraje*
Cotzal	Huipil	Sash	....	*Perraje*
Chiché	Huipil	....	....	*Perraje*
Chichicastenango	Huipil	Belt	....	*Perraje*
Nebaj	Huipil	Belt	Headdress	*Perraje*
Zacualpa	Huipil	Sash	Headdress	*Perraje*

Joyabaj is another *municipio* where backstrap weaving is done. Some men there still dress in the traditional costume. A few men from Chichicastenango also wear their old costumes. In general, however, the sash is the only item in the men's costume that is still woven on the backstrap loom. Costumes in Chiché and Chichicastenango are the same. Wool belts in Chichicastenango are made by specialists. Chiché women get their belts from Chichicastenango.

Department of Sololá

There are a number of *municipios* around Lake Atitlán where backstrap weaving is done and the men wear traditional costumes. We visited only two towns in the department: Nahualá and Santa Catarina Ixtahuacán. Indians in these two towns dress alike. Women's huipils, sashes, and *perrajes* and men's shirts, pants, sashes, and *tzutes* are woven on the backstrap loom. Virtually all of the women do weaving.

Department of Sacatepéquez

We visited only San Antonio Aguas Calientes in this department. Women there weave huipils, *perrajes*, and belts for themselves and for sale. San Antonio weavers know how to weave in different styles and techniques, and they have adapted European embroidery patterns to backstrap weaving. In the past few years San Antonio has become a tourist attraction, and it remains to be seen how the demand for cheap handwoven souvenirs is going to influence the local standards of weaving, which up to now have been very high. Santo Domingo Xenacoj, Santa Maria de Jesús, San Lucas Sacatepéquez, and Santiago Sacatepéquez are some of the other *municipios* in this department where backstrap weaving is done.

Other Departments

In the departments of Alta Verapaz and Baja Verapaz backstrap weaving is on the decline. An exception is the *municipio* of Tactic (Alta Verapaz), where women weave huipils that are sold in many towns. In the department of Chimaltenango weaving is done in several *municipios.* Of special interest are the textiles from Comalapa. In the department of Escuintla weaving is done by a few women in Palin. San Juan Sacatepéquez and San Pedro Sacatepéquez are two of the *municipios* in the department of Guatemala where backstrap weaving is done. In the department of San Marcos, the town of San Pedro Sacatepéquez used to be an important center for backstrap weaving. Since we did not get to this area, we do not know to what extent backstrap weaving is still done there.

GUATEMALAN BACKSTRAP WEAVING

Plate 1. Beating raw cotton in Todos Santos Cuchumatán.

1

WEAVING MATERIALS

Before the arrival of the Spaniards yarn was spun from cotton, maguey, and other plant fibers. Cotton yarn is still the most important material for the backstrap weaver. After the Spaniards introduced sheep, the Indians learned how to work with wool. Silk from China reached Guatemala during colonial times and was used for embroidery and extra weft patterns. Toward the end of the nineteenth century commercial cotton yarn began to replace homespun cotton. Within the last fifteen years acrylic yarn has become popular as a pattern yarn.

HAND-SPUN YARN

While most weavers depend on commercial yarn, hand-spinning is still practiced in a number of places where the raw material—wool, maguey, or cotton—is locally produced or easily available.

Cotton

We observed spinning of cotton in Todos Santos, San Juan Atitán, San Sebastián Huehuetenango, and Chiché, and we assume that a few women in other places also spin cotton occasionally. Only in Todos Santos is spinning still fairly common. People from this *municipio* work every year on the cotton plantations on the coast and bring back sacks of raw cotton. The cotton is sold on the market for fifteen centavos per pound. Only white cotton is available.

Fiber Preparation Observed in Todos Santos

The fibers are separated from the seeds and loosened with the fingers. After approximately four ounces of cotton is prepared in this way, a bundle of rags is placed on the ground and covered with a goatskin, hairy side down. The skin is held down at one end by a rock. The cleaned cotton is placed on the goatskin. The woman sits on her heels in front of the goatskin, holding it down with her knees. She uses two forked sticks alternately to beat the cotton (see plate 1). She holds the sticks at an angle of somewhat less than 90 degrees. The pile of cotton is beaten vigorously at a rate of about 4 beats a second until it has spread out into an elliptical sheet about 30 inches long at the longer

Plate 2. Folding the beaten cotton in Todos Santos Cuchumatán.

Plate 3. Folding the cotton into a strip in Todos Santos Cuchumatán.

axis. The woman now folds the sheet in half along this axis and beats it again until its size is about 20 × 60 inches. Again, she folds it along the longer axis and then rolls it up. The roll is placed in front of the goatskin, and one end is put on the skin. Now, the whole length of cotton is beaten and moved forward in the process. It is folded (see plate 2) for the third time and rolled up. The woman now places the roll to the right of the goatskin and moves the band of cotton to the left while beating it for the fourth time. She then folds it into a strip 3 inches wide and rolls the strip into a ball (see plate 3). For spinning, about 6 inches of cotton is pulled from this strip.

Spinning As Observed in Todos Santos

Spindles are straight, slender sticks, about 15 inches long. They have pointed ends and a clay whorl a few inches above the lower end. Spindles are sold in the market by itinerant merchants.

The spinner sits on her heels and holds a short strip of unspun cotton in her left hand. The upper end of the spindle is moistened and the cotton is attached to it. The tip of the spindle is set in a gourd or cup. While the spindle is twirled with the index finger and thumb of the right hand, the left hand draws out the cotton to form a thread. The thread is held up during the spinning so that the angle between spindle and thread is much larger than 90 degrees. The spindle is twirled about 14 times, until 30 to 40 inches of yarn has been formed. In some sections the yarn is too thick and has to be thinned by stretching it with the fingers. Next the yarn is wound on the spindle. During this procedure the angle between spindle and thread is close to 90 degrees. Plates 6 and 7 show the two different positions of spindle and thread for spinning and winding. It takes about fifteen seconds to spin and wind 35 inches of yarn. Some weavers leave the spun cotton on the spindle, using the spindle as a bobbin. Otherwise, the cotton thread is wound into a ball.

Nowadays, handspun cotton is used mainly for the weft where it does not show.

Plate 4. Spinning cotton in San Sebastián Huehuetenango. The cotton was prepared with the cards lying in front of the woman.

Fiber Preparation Observed in San Sebastián Huehuetenango

After taking out the seeds and fluffing up the fibers with the fingers, the weaver cards the cotton with wire brushes (cards) in the same way that wool is carded. The result of carding is fluffy rolls, about 5 inches long. Spinning is the same as in Todos Santos (see plate 4). Carding the cotton seems to be faster and less strenuous than beating it.

Wool

In the department of Huehuetenango many weavers use hand-spun wool yarn to make belts, blankets, skirts, and many other items. People in San Sebastián Huehuetenango told us that thirty or forty years ago most people wore only woolen clothing. In some *aldeas* of San Sebastián, where people have many sheep, women wear woolen huipils, skirts, belts, and headbands, and the men's shirts and *delanteras* are also made from wool yarn. All these items are woven on the backstrap loom.

Raw wool is sold on the market for about sixty

Plate 5. Carding wool in San Sebastián Huehuetenango.

centavos a pound. It is available in black (which fades into brown) and white. To obtain shades of gray, black and white wool are blended. Wool is washed in lukewarm water, dried in the sun, and carded with wire brushes (plate 5). Spines and seeds that remain after carding have to be removed by hand. To obtain a fine yarn, wool is spun with a spindle, just like cotton (plates 6 and 7). For a coarser yarn a spinning reel is used.

Maguey

In several *municipios* maguey is grown to provide the raw material for making twine, rope, nets, straps, and bags. Maguey fiber is not used for weaving clothes.

COMMERCIAL YARN

Most of the commercial yarns are made in Guatemala. They are sold on the market and in stores. Prices vary from one town to the next and are higher in remote areas.

Single-Ply Cotton Yarn

Single yarn is the least expensive and therefore the most popular yarn for weaving. It is known as *hilo* or *hilo flojo*, that is weak yarn. For use as a warp it has to be twisted to make it stronger, or else two or three singles are combined. Often two plies are twisted together. Single yarn comes in different sizes designated by numbers. The coarsest yarn used for weaving is no. 8; no. 20 is the finest. The numbers indicate how many hanks of a standard length make up one pound of yarn.

The cheapest and most widely used cotton yarn is unbleached white. It has a creamy color and costs about one quetzal per pound (the quetzal has the same value as the dollar). In Todos Santos and Colotenango, no. 8 of this yarn is used as single for warps and wefts. In San Juan Atitán, no. 12 is paired in warp and weft. Bleached white yarn, called *hilo chino*, is more expensive (1.25 quetzals or more per pound). It is used in Nahualá (nos. 14 or 16 paired in warp and weft), Nebaj and Chajul (no. 20 in pairs or triplets for warp and weft).

Dyed single yarn is available in many colors, but only red and dark blue yarn are of importance to weavers. When other colors are needed, two-ply yarn is used in most cases. The cheapest of the red yarns costs 1.60 quetzals per pound. It is not colorfast and fades quickly to a pink. Most weavers use it only for the weft where it does not show

Plate 6. Spinning wool in San Sebastián Huehuetenango.

Plate 7. Spinning wool in San Sebastián Huehuetenango. The woman is about to wind the spun yarn onto the spindle.

much. A better grade of red yarn costs about 2.20 quetzals per pound. More expensive and hard to get is the so-called *hilo Alemán*, an imported yarn that is known for its colorfastness. Good-quality dark blue yarn for skirts is very expensive too. During our stay in Guatemala the price of this yarn rose from 2 to more than 3 quetzals per pound.

In Zunil, Chiché, and Chichicastenango, weavers use tie-dyed yarn from Salcajá for carrying cloths and *servilletas*.

Two-Ply Cotton Yarn

Two-ply cotton yarn bearing the brand name Mish does not need twisting or doubling and has more luster than the single yarns on the market. It shrinks less than the single yarns. Two-ply Mish comes in about twenty colors. Its size is comparable to two no. 20 singles that have been twisted together. If it were cheaper, two-ply yarn would be the most popular yarn for weaving. Two-ply yarn costs a minimum of 2.40 quetzals per pound. One pound consists of twenty hanks. A hank costs at least 12 centavos, and one-eighth of a hank is available for 2 centavos. Many weavers use two-ply Mish for plain weaving and for brocading.

Three-Ply Cotton Yarn

Three-ply Mish has the same qualities as two-ply Mish, but it is stronger. It costs 2.40 quetzals per pound. Three-ply yarn is used by weavers in San Antonio Aguas Calientes to weave huipils. In Nebaj three-ply yarn is used to make belts. In Aguacatán we found headbands made from three-ply Mish. In Zacualpa a special kind of three-ply yarn is used as a supplementary weft. It is a rich purplish-blue color, and merchants told us that it is dyed twice for better colorfastness. One pound costs 5.60 quetzals, and one ounce costs 35 centavos.

Mercerized Cotton

Mercerized cottons for brocading are imported and rather expensive. They are known as Sedalina or Lustrina. (In the Huehuetenango area, where most weavers cannot afford to buy expensive yarns, the name Lustrina is applied to two-ply Mish.)

Sedalina, a two-ply yarn from Colombia, costs twenty centavos per five-gram ball. Other brocading yarns from Spain, Mexico, and Eastern European countries are sold in extremely small hanks for six to eight centavos per hank.

Acrylic Wool (Acrilan)

Acrylic wool, a synthetic fiber, has in recent years become popular as a supplementary weft. Occasionally it is used for plain weaving too. Acrylic wool is much cheaper than mercerized cotton (3.20 quetzals per pound), and its garish colors are well liked.

Rayon

Rayon is used as brocading yarn in the *municipios* of Nahualá and Santa Catarina Ixtahuacán. Most popular is a purple yarn (of the brand Visco Suisse) whose color bleeds during washing. Bleeding is caused deliberately by soaking newly woven textiles overnight in soapy water. Hanks of rayon (one ounce?) cost from sixty centavos to one quetzal each.

Silk

Silk was once widely used for brocading. Today the price of silk floss is prohibitive: forty to fifty centavos for a quarter ounce. Weavers in Chajul still use silk for ceremonial huipils; otherwise silk has been replaced by less expensive yarns.

Commercial Wool Yarn

In the past some weavers used commercial wool yarn for extra-weft patterns. Today commercial wool has been replaced by Acrilan, an acrylic yarn that resembles wool.

Cost of Yarn

When Lila O'Neale studied Guatemalan textiles in 1936, the price of the cheapest commercial yarn (unbleached white cotton yarn) was thirty centavos per pound, and colored yarn cost twice as much. The daily wage of a plantation worker was thirteen centavos. In 1976, Indians working on plantations earned about eighty centavos per day, and a pound of white cotton yarn cost one quetzal. Thus yarn was cheaper in 1976 than in 1936, but buying yarn is still a big expenditure for most weavers. Many women use the more expensive two-ply yarn only for the warps and the cheaper single yarn for the weft.

Plate 8. Twisting commercial yarn with a spinning reel in San Sebastián Huehuetenango. This kind of reel is used for spinning wool and for twisting cotton yarn.

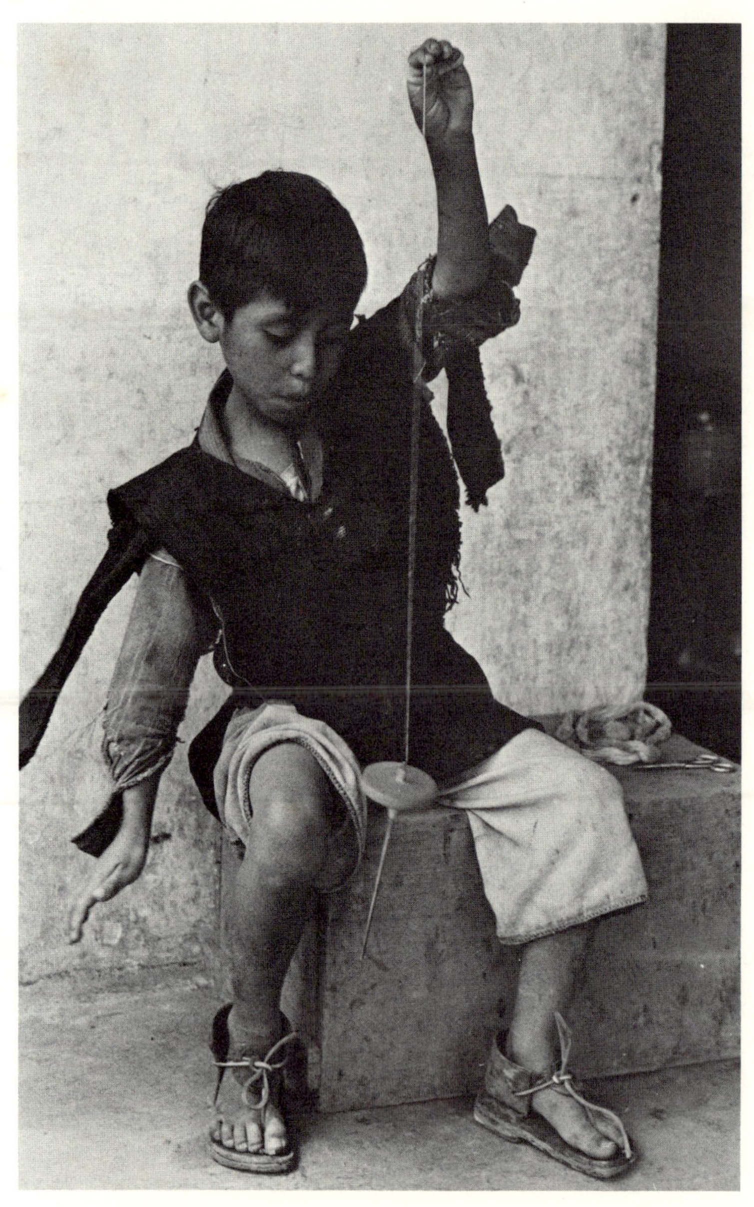

Plate 9. Twisting several strands of yarn together with a suspended spindle in San Juan Atitán.

Fig. 1. Doubling wool yarn with a spindle as seen in San Sebastián Huehuetenango.

2

TWISTING AND DOUBLING

To give more strength to a factory-made single yarn, it is often twisted with a spindle or spinning reel (plate 8). Often two singles are twisted together into a two-ply yarn. Twisted yarn is used mainly for the warp.

Twisting and doubling is common practice in the Huehuetenango area. We observed it in San Sebastián Huehuetenango: two hanks of a single yarn are placed on a reel and wound together into a ball. Then the paired yarn is twisted with a spindle. The woman holds the ball up high with her left hand while she twirls the spindle about ten times with her right hand. After that, the twist is checked by decreasing the tension of the yarn. The twisted length is wound on the spindle. It takes about fifteen seconds to twist and wind thirty-five inches of yarn.

For the doubling of handspun wool, a different technique is employed. Yarn from two spindles is wound into a ball and twisted together with a spindle that is held in a horizontal position. The spindle is held at one end by the fingers, and at the other end it is suspended by the yarn (figure 1). This way of twisting is rather slow. Doubled wool yarn is used for belts.

In Todos Santos and San Juan Atitán men crochet their own bags. For this they need a strong thread, which they get by twisting several plies of yarn together. They use a suspended spindle (plate 9) for twisting. We saw only men using this spindle.

3

»»»»»X«««««

DYEING

Commercial cotton and synthetic yarns are available in all colors. They are dyed in the factory or by specialists (tie-dyeing). The backstrap weaver who works with these yarns does not need to do any dyeing.

In the Huehuetenango area where red wool is used for headbands and belts, a number of people know how to dye wool. Hanks of handspun wool yarn or finished strips of material for headbands are dyed with aniline dyes. We observed the dyeing of headbands in San Sebastián Huehuetenango: the men who weave the headbands (on a simple apparatus that combines features of both the treadle and the backstrap looms) also dye them after weaving. The long strip of woven wool material to be dyed is put in a clay jar with hot soapy water, then placed on a board and worked over with the bare feet. This procedure is repeated several times. Then the strip is rinsed and boiled in another clay jar that contains dye. Two kinds of aniline dye are combined to produce the right shade of red. Lemon juice is used as a mordant. From time to time the color of the strip is checked, and more dye is added to the mixture if necessary (plate 10). This continues until the right color has been reached. Then the band is hung up to dry. Hanks of handspun wool yarn are dyed before weaving using the same method.

Plate 10. Dyeing woven wool strips for headbands in San Sebastián Huehuetenango.

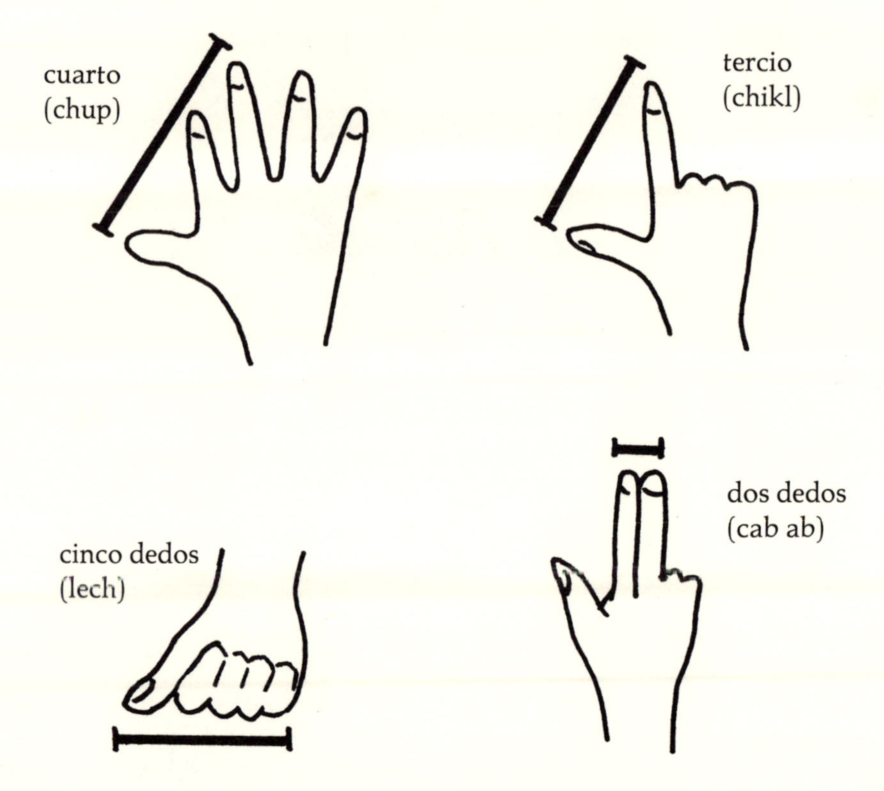

Fig. 2. Units of measurement used in San Sebastián Huehuetenango. The names of the measurements are given in Spanish and, in parentheses, in Mam.

4

UNITS OF MEASUREMENT

Warping and setting up the loom, which will be discussed in following sections, require measurements. The weaver uses her hand and fingers as a measuring device. Figure 2 shows various units of measurement that are used in Sujal, an *aldea* of San Sebastián Huehuetenango. The *cuarto*, a unit of measurement about 7 inches long, and the *dedo* (the width of a finger) are used most often. The measurements of a piece of material are given as so many *cuartos* plus so many *dedos* (*cuarto* and *dedo* are Spanish translations of the Indian words).

In some places, the distance between the outstretched thumb and middle finger is called *cuarto*.

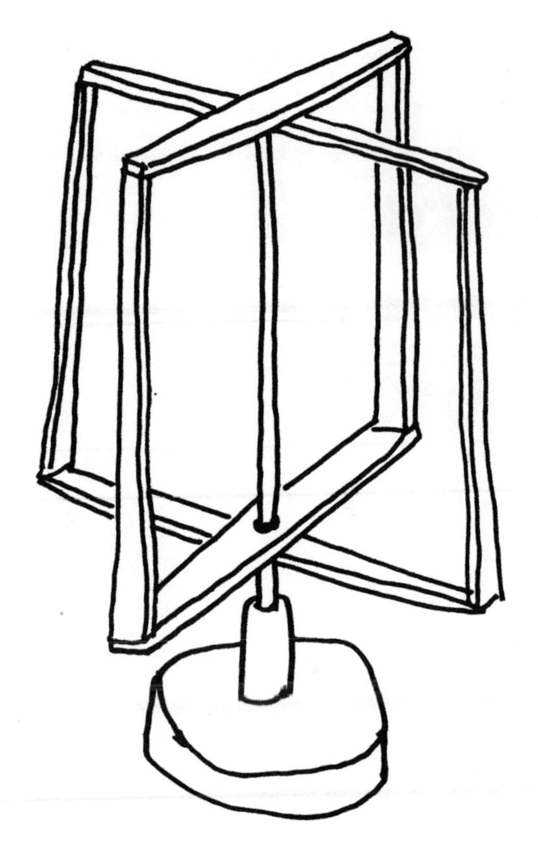

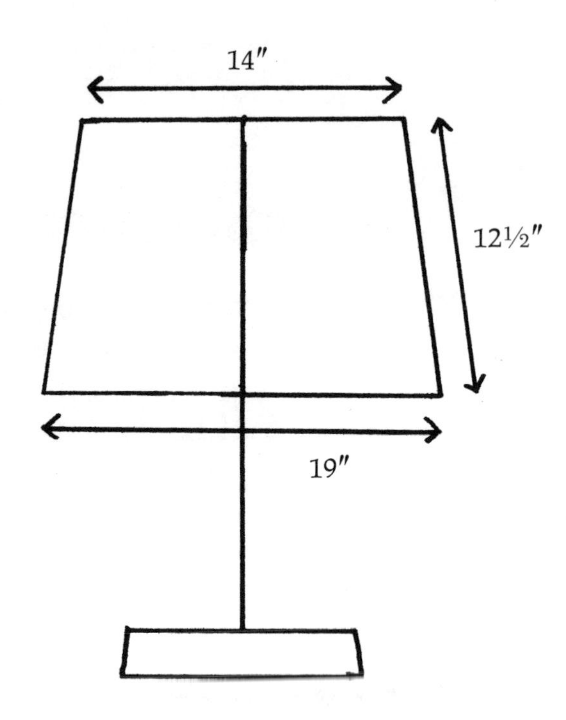

Fig. 3. Reel from Todos Santos Cuchumatán.

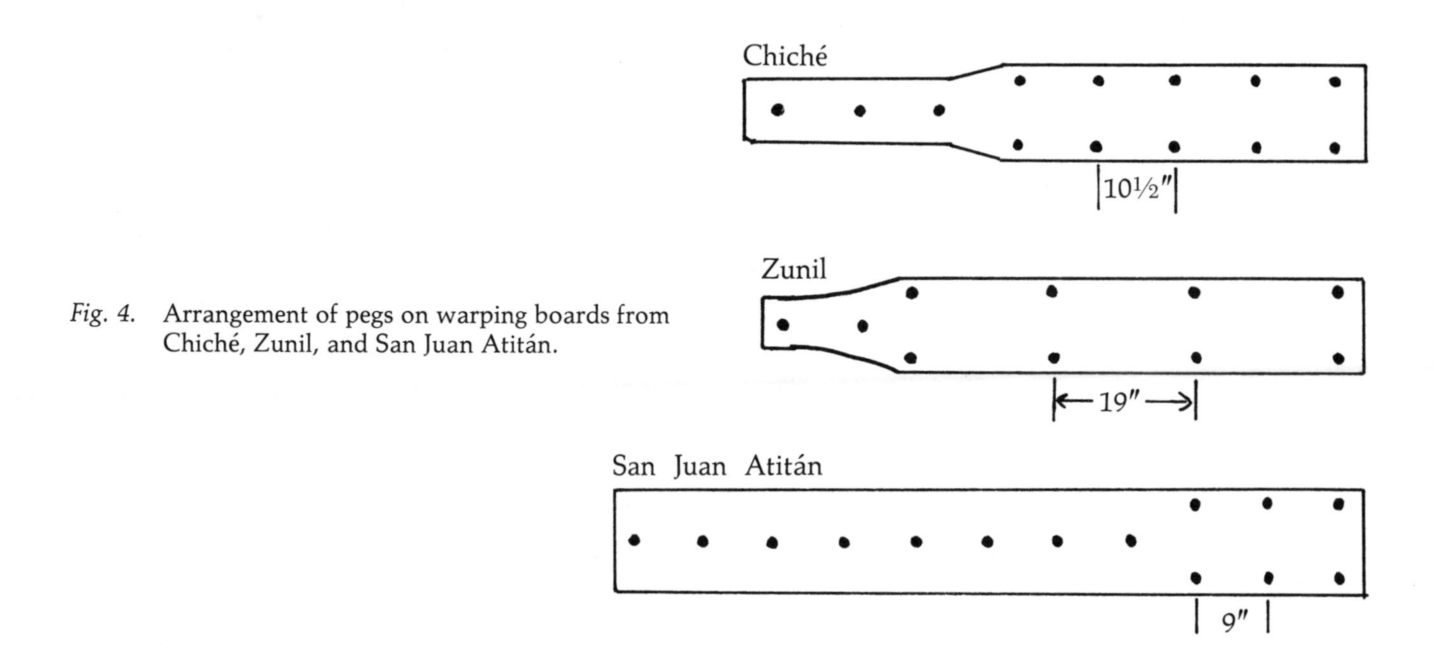

Fig. 4. Arrangement of pegs on warping boards from Chiché, Zunil, and San Juan Atitán.

5

»»»»»»»»»»«««««««««

WARPING

To arrange warp yarns for the loom, the weaver winds yarn around the pegs of a warping board, or around sticks that are driven into the ground. The distance from the first to the last warping peg is equal to the length of the warps when they are placed on the loom. The finished piece, however, will be somewhat shorter because of take-up (chapter 12). Weavers take this into account and make the warps slightly longer than the required length of the finished piece.

Warping divides the yarn into two groups of warp elements, the odd- and the even-numbered warps. On the loom these two groups of warps will be separated by the shed roll. Warp crosses keep the warps fixed in the same sequence in which they were wound around the pegs of the warping board. All the weavers that we observed arranged the warps in two crosses.

TOOLS FOR WARPING

Commercial yarn is sold in hanks. To unwind the yarn from the hank, it is necessary to have a reel. Reels are often homemade and look much the same everywhere (figure 3 and plate 11). From the reel the weaver winds the yarn directly onto the pegs of the warping board, or else she winds the yarn into balls. The first method is employed when warps of only one color are needed. If many colors of yarn are used for the warps, it is more convenient to wind the yarn into balls, each ball a different color. Then the yarn is wound around the pegs of the warping board.

In many places weavers combine two or three single yarns to get a stronger warp. This is done by placing two or three hanks of yarn on one reel and unwinding them together.

Homespun or hand-twisted yarn is often left on the spindle. From the spindle it is wound on the pegs of the warping board, as shown in plate 14.

Warping boards differ in size, number, and spacing of pegs. They are made by carpenters. In San Sebastián Huehuetenango only a few weavers own a warping board. Weavers who do not have warping boards hammer sticks into the ground for warping. The warping procedure is the same for both. Warping on sticks is shown in plates 12 and 13;

Plate 11. Winding commercial yarn into a ball in Todos Santos Cuchumatán.

a warping board can be seen in plate 14. In figure 4 we can see the arrangement of pegs on warping boards from three different villages. Warping boards are 4 to 8 feet long. The longest warps—up to 12 feet long—are required for belts and sashes and for weaving two panels of a huipil in one setup.

BASIC WARPING PROCEDURE

Let us assume that warping is done on three sticks with yarn of one color. The weaver kneels or sits on her heels in front of the warping sticks and unwinds yarn from a reel. At the beginning of the warp, a loop is made which goes around stick 1 (figure 5). From stick 1 the warp travels to the left and passes in front of stick 2, then goes around stick 3 and returns to stick 1. This way, two warp crosses are formed between the sticks. The circuit from stick 1 to stick 3 and back to stick 1 is repeated until the required number of warps has been wound on the sticks. The last warp ends with a loop that is placed over stick 1.

Each complete circuit of yarn is counted as one by the weaver. The weaver uses a string as a count-

Plate 12. Warping yarn for a *servilleta* on three stakes in San Sebastián Huehuetenango.

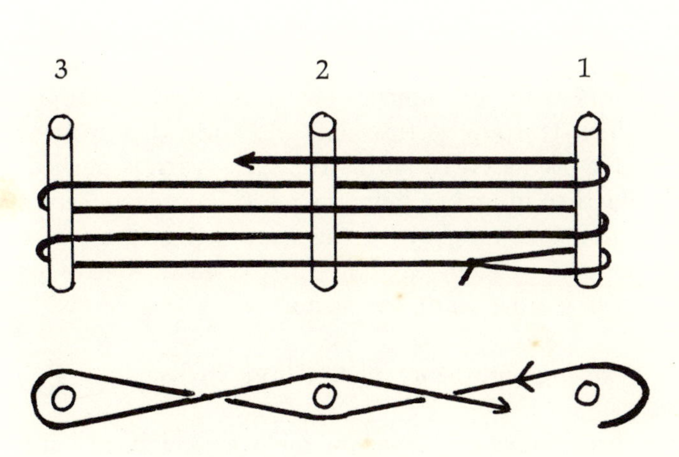

Fig. 5. Warping on three sticks. Side view and view from above.

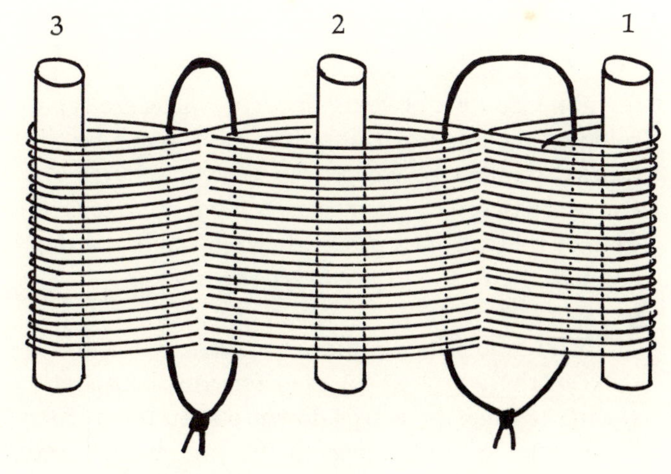

Fig. 6. Tying the warp crosses.

Plate 13. Warping yarn in San Sebastián Huehuetenango for a *delantera*. Black and white wool are used alternately in a *cadena* technique.

ing aid. Groups of twenty or forty warps are separated or encircled by this string. After reaching the desired number of warps, the weaver secures the warp crosses with strings as shown in figure 6. One of these strings passes through the loops that connect the first and last warps to stick 1.

Throughout the warping, the weaver is careful to keep the warps at an even tension. Frequently she has to push the warps down to keep them from "climbing" up on the pegs. If the yarn breaks and warps have to be tied together, the knots are made close to stick 1. After the loom is set up, these knots will not interfere with the heddle loops. When warping is finished, and the warp crosses have been secured with strings, the yarn is pushed together and taken from the pegs. To prevent separation and tangling of the warp yarns, the warp skein is twisted, as if to turn it inside out. The warp is now ready for sizing.

Warping with More Than Three Sticks

Three sticks, or pegs, on the warping board are used for warping when the warp is fairly short. In most instances, however, weavers use five or

Plate 14. Warping on a warping board in Todos Santos Cuchumatán. The warp yarn comes from a spindle.

more pegs. In figure 7 different ways of warping are shown. There are two warp crosses in each of these warp setups.

WARPING STRIPES

The weaver has a sample at hand in which she counts the number of warps that go into one stripe. Yarn for the stripes is wound into balls with the help of a reel. Warping starts as usual at stick 1. After finishing the warping of the first stripe, the weaver breaks the yarn and ties yarn of a different color to it. All knots are made at the same place in the warp circuit where the yarn goes around stick 1. The weaver does not need a string as a counting aid since the stripes themselves serve as markers.

If the material to be woven has a great number of stripes, it would be very tedious to tie the warps together with knots. It is more convenient to warp the stripes using a *cadena* (chain) technique. Most weavers are familiar with at least one way of doing the *cadena.* Common to all methods is the principle of having warp yarn of one color cross over from one warp stripe to the next.

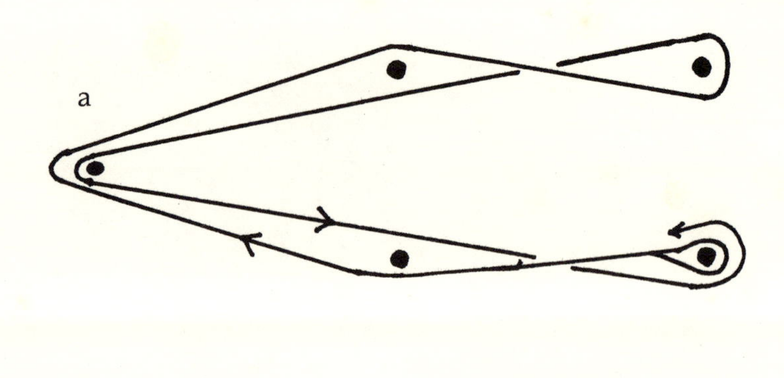

Fig. 7.
Warping on 5, 7, and 10 pegs: *a,* San Sebastián Huehuetenango, yarn for one panel of a huipil; *b,* Zunil, yarn for one panel of a huipil; *c,* San Martín Sacatepéquez, yarn for two panels of a huipil.

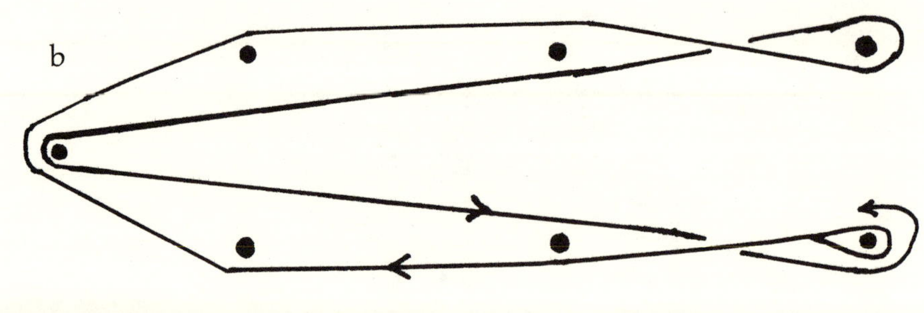

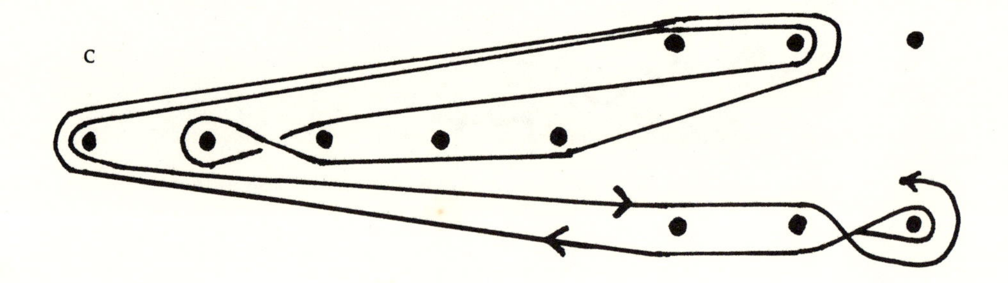

In San Sebastián Huehuetenango we observed a weaver warping a *servilleta.* She used red, white, and orange yarn in turns, employing a *cadena* technique for the red and white yarn only. Warping was done on three sticks. In the following description we explain the basic procedure that was followed by the weaver. For simplicity we shall assume that only red and white yarn was used:

The weaver started the white yarn at stick 1, carried it past stick 2, around stick 3 and back to stick 1, as in regular warping. The second circuit of the white yarn was done in the same way. However, after passing stick 2 on the way back, the yarn was placed on the ground in front of the weaver. The second circuit of white yarn was left incomplete. The weaver placed a rock on top of the white yarn to keep it under tension, and then she started a red warp on stick 1. With the red yarn she completed one circuit and started a second one. Before the second circuit was complete, she placed the red yarn on the ground and picked up the white yarn to continue the warping. Throughout the warping she used red and white yarn in turns.

Figure 8 shows the warping sticks as seen from

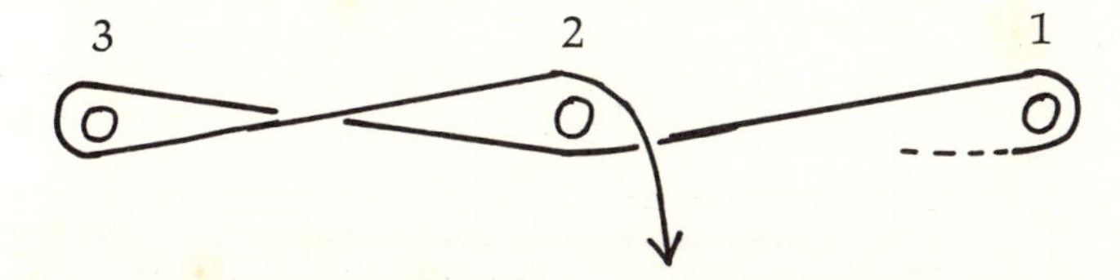

Fig. 8. *Cadena,* San Sebastián Huehuetenango.

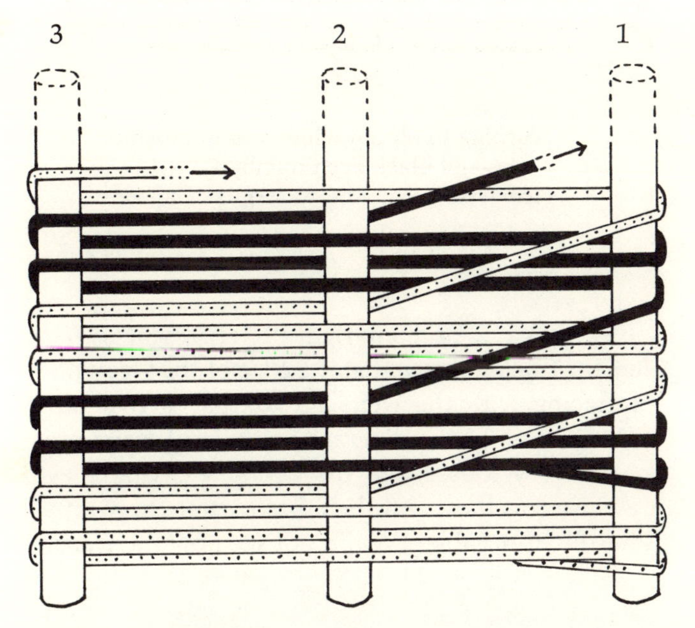

Fig. 9. Warping stripes in a *cadena* technique, San Sebastián Huehuetenango.

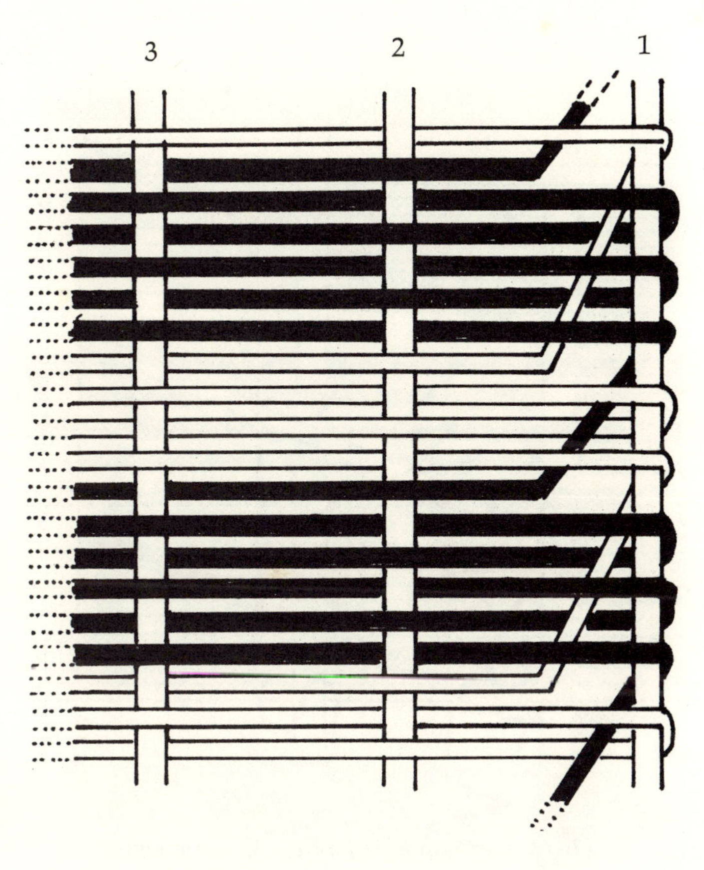

Fig. 10. *Cadena* on a *tzute* from San Juan Atitán.

above. Where the warp yarn is placed on the ground, it is shown ending in an arrow.

Figure 9 shows the crossover of the white and red warps on the warping sticks. As in standard warping, there are two warp crosses. When warping is finished, the warp crosses are secured with strings. Later on, when the weaver transfers the warps to the loom sticks, she takes great care to arrange the stripe pattern in the same way that it appeared on the warping sticks.

On the finished piece the crossover is visible at one of the end selvages. The first and second weft at the edge of the weaving are, in relation to the warps, positioned the same as sticks 1 and 2 in figure 9. If sloppily done, this kind of crossover is easily noticed because of loose warp threads at the edge of the weaving.

The *cadena* technique described above is used by weavers in San Sebastián. In other towns weavers employ different kinds of *cadenas.* Figure 10 shows how the crossover appears at the edge of a woven piece from San Juan Atitán. Weavers in Colotenango and San Ildefonso Ixtahuacán also use this method.

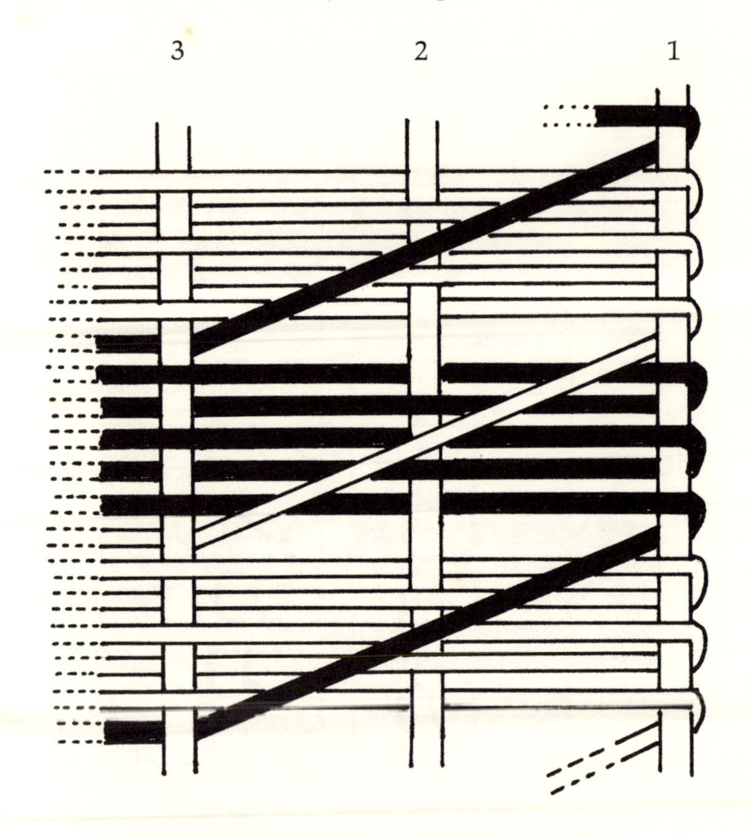

Fig. 11. *Cadena* on a huipil from Colotenango.

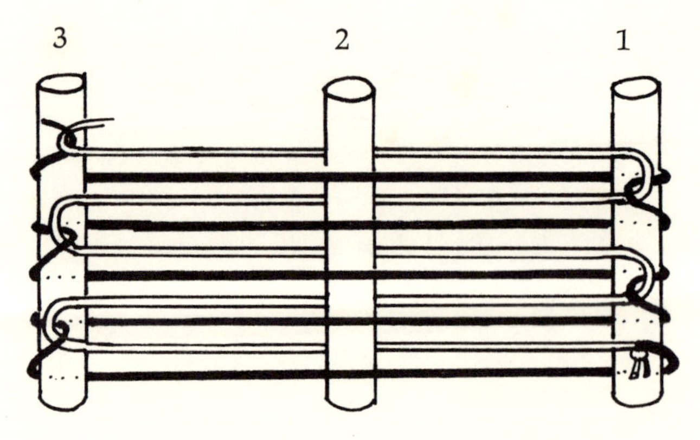

Fig. 12. Warping to obtain white odd-numbered warps and black even-numbered warps (or vice versa), San Antonio Aguas Calientes.

Elements 1, 2, and 3 in figure 10 represent *wefts*, element 1 being the weft at the edge of the weaving. Interlacing with the wefts are warps of two different colors. In the diagram warps and wefts are spaced very loosely. In the actual weaving they are close together, and the crossover of the warps is hardly noticeable. We did not see how the warping was done, but it can be assumed that wefts 1 and 2 correspond to pegs on the warping board.

On textiles from Colotenango we found yet another kind of *cadena*. Figure 11 shows how the crossover appears at the edge of the weaving. Again, elements 1, 2, and 3 represent wefts. This kind of crossover shows on only one side of the material. When the material is made into a huipil, the crossover is on the inside of the garment. Both this kind of *cadena* and the one from San Juan Atitán are more elegant than the *cadena* from San Sebastián Huehuetenango.

In the Huehuetenango area many stripe patterns are done in three colors—usually white, red, and orange. Often all three colors are warped in a *cadena* technique.

ODD- AND EVEN-NUMBERED WARPS OF DIFFERENT COLORS

During the warping the weaver counts each complete circuit of the yarn on the warping board as one. Once the warps are arranged on the loom, however, each warp element going from one end bar to the other is counted as one. This means that there are twice as many warps on the loom as there were warp circuits on the warping board. The shed roll divides the warps into two groups. For convenience we shall call the warps that pass over the shed roll "odd-numbered warps" (warps 1, 3, 5, and so forth). Warps passing under the shed roll are the even-numbered warps (numbered 2, 4, 6, and so forth).

For some stripe patterns, and for the weaving of warp-float patterns, it is necessary for the even-numbered warps to be of a different color from the odd-numbered ones. This requires a warping procedure that is slightly different from the methods discussed so far. Let us assume that black yarn alternates with white yarn in a belt and that warping is done on three pegs. The white yarn travels from peg 1 to peg 3. At peg 3 it is broken and the black yarn is tied to it. The black yarn travels back to peg 1, and so on. White and black yarns always travel on different paths. On the loom they are separated by the shed roll. Instead of tying the warps together, most weavers use a *cadena* technique, as shown in figure 12.

Plate 15.

Sizing a warp skein in a corn-dough solution at San Juan Atitán.

Plate 16.

After sizing, the San Juan weaver spreads the warps over the end bars. She is beating the warps with a stick to free them from particles that were in the sizing solution. Her daughter is stretching the warps.

6

»»»»»»«««««

SIZING

Sizing is the next step after warping. Sizing makes the warp yarns stiffer, stronger, and smoother. In virtually all the places that we visited, sizing was done; the only exception was San Antonio Aguas Calientes. There strong three-ply yarn is used for the warp, and sizing is regarded as unnecessary.

To make a sizing solution, corn dough is dissolved in hot water. Weavers know from experience how much dough is needed for proper sizing. Too much starch makes the warp thick and the weaving coarse.

If the weaver owns a sieve, she passes the corn dough solution through it. Then the warp skein is soaked in the solution (plate 15) and kneaded. The skein is wrung out and left to dry while tightly stretched between two loom sticks. To get rid of corn particles that stick to the warps, the weaver beats the stretched warps with a stick (plate 16).

Sizing causes the yarn to shrink. Commercial one-ply yarn shrinks approximately 1.5 percent; the more recently developed two-ply yarn shrinks less than 0.5 percent. If the two kinds of yarn are used together, two-ply warps will be noticeably longer than one-ply warps after sizing. To correct this defect, after the first few wefts have been put in, two-ply warps are pulled beyond the edge of the weaving with a needle to take up the slack and even out the warps.

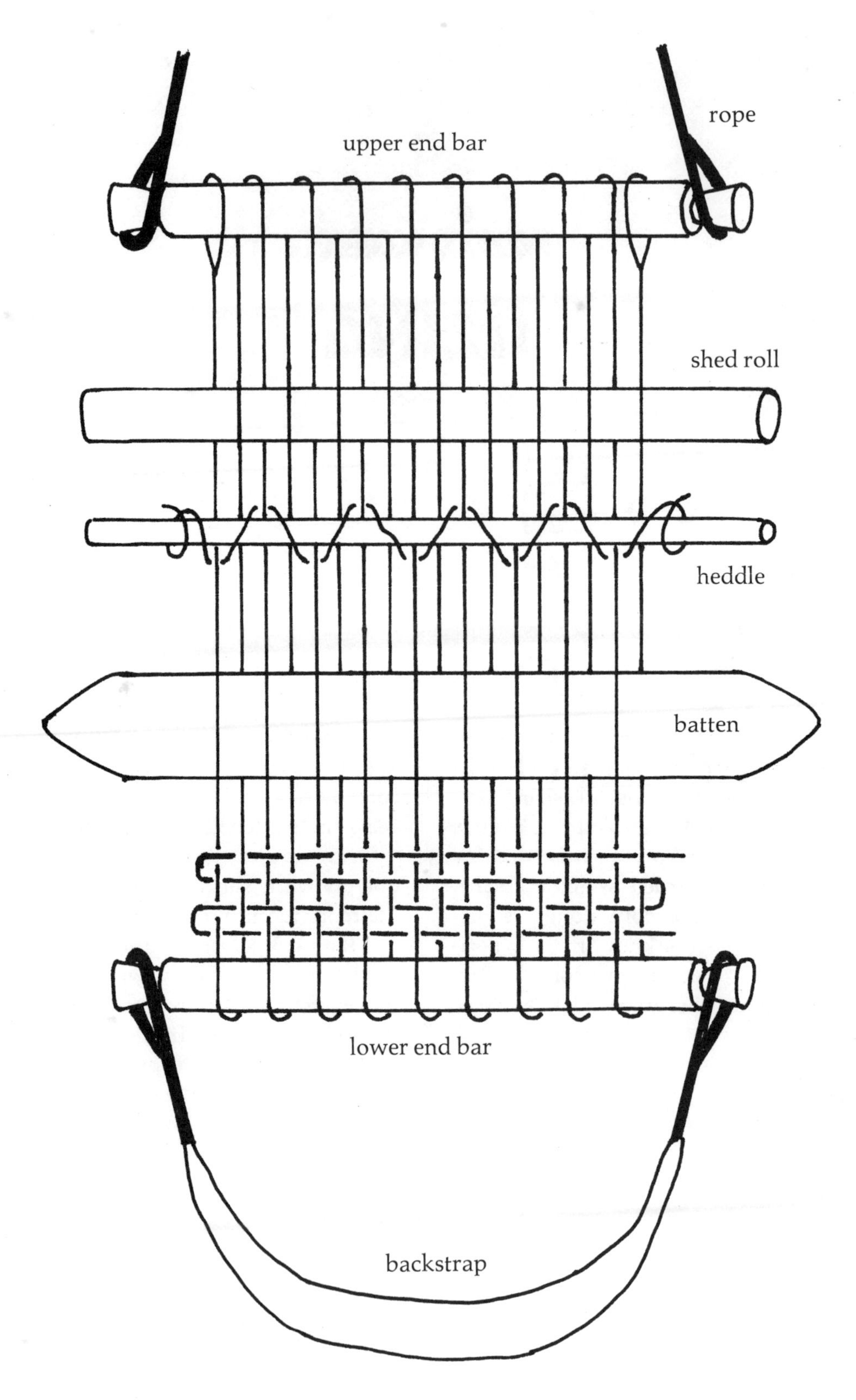

Fig. 13. Loom for weaving with end fringes, basic parts.

7

»»»»»»»»»»»»»»»«««««««««««««««

THE BACKSTRAP LOOM: LOOM PARTS

The backstrap loom is assembled from a few sticks, rope, and string. When not in use it takes up very little space and can be easily transported or stored.

Two *end bars,* a *rope,* and a *backstrap* are needed to tension the warps. For weaving with end selvages, the warps are tied to the end bars with strong cords *(loom strings).* To establish the shed and the countershed for plain weaving, the loom is equipped with a *shed roll* and a *heddle.* The *batten* helps to separate odd- and even-numbered warps, keeps the sheds open, and battens down wefts and warp crosses. Weft picks are made with a *bobbin,* the width of the fabric is controlled by a *tenter,* and a *cloth bar* is employed to roll up the woven material. In addition to these parts, many looms are equipped with *lease sticks.* They make weaving easier, but are not absolutely necessary. For weaving techniques other than plain weaving and for adding extra wefts to the ground fabric, additional loom parts may be necessary. They are described in chapters 18 and 19.

THE ROPE

A strong rope, with a loop at each end, serves to hitch the upper end bar of the loom to a house post, rafter, or tree. Care is taken to tie the rope so that both ends of it are the same length and the upper end bar rests in a horizontal position.

THE END BARS

The warp yarns are tightly stretched between the upper and the lower end bars. The upper end bar is connected to a rope, which is fastened to a house post, rafter, or tree. The lower end bar is attached to the backstrap, which goes around the weaver's hips. The weaver regulates the tension of the warps by leaning backwards or forwards.

End bars are smoothly rounded and have notched ends. Some weavers use homemade end bars, others buy them from a carpenter. Occasionally end bars

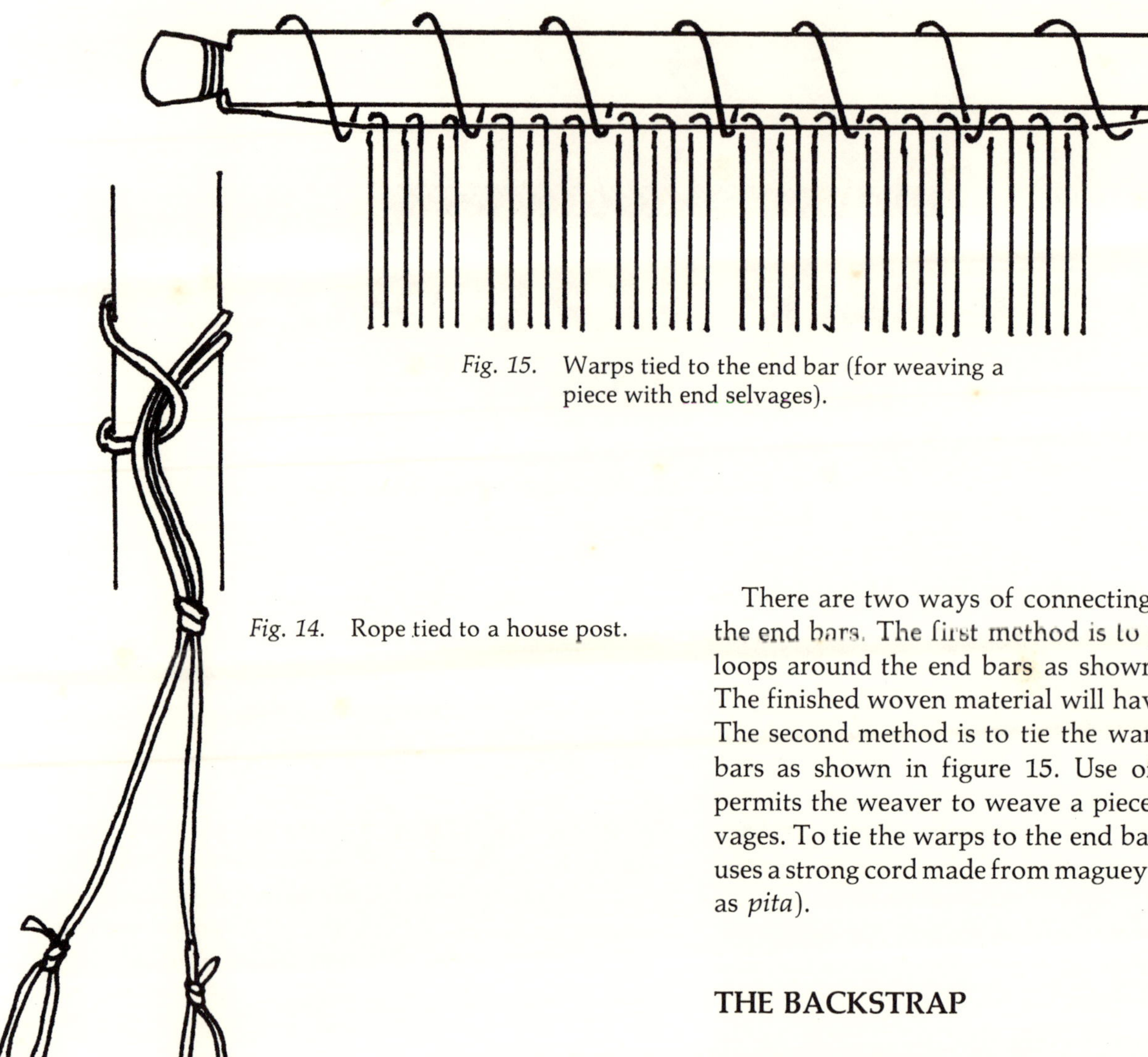

Fig. 15. Warps tied to the end bar (for weaving a piece with end selvages).

Fig. 14. Rope tied to a house post.

are sold on the market. In Nahualá a set of three medium-sized bars (32 inches long) costs forty centavos. End bars are from 15 inches to 45 inches long, depending on the width of material to be woven. In San Sebastián Huehuetenango, many weavers have three sets of sticks. The largest one, for weaving skirts, measures about 38 inches; a medium-sized set (32 inches) is for huipils and *tzutes*, and the smallest set—18 inches—is for weaving belts and sashes. End bars are made from pine *(pino blanco)* or from more expensive hardwood.

There are two ways of connecting the warps to the end bars. The first method is to pass the warp loops around the end bars as shown in figure 13. The finished woven material will have end fringes. The second method is to tie the warps to the end bars as shown in figure 15. Use of this method permits the weaver to weave a piece with end selvages. To tie the warps to the end bars, the weaver uses a strong cord made from maguey fibers (known as *pita*).

THE BACKSTRAP

The backstrap is fastened by short ropes to the lower end bar. The weaver stretches the warps tight by leaning back against the backstrap, which passes around her hips. Most backstraps are made from maguey-fiber cord and cost about fifteen centavos on the market. In Nahualá, Chiché, Aguacatán, and San Antonio Aguas Calientes we saw backstraps that were made from leather.

THE SHED ROLL

The shed roll separates the odd- from the even-numbered warps, producing *shed 1*, as we shall call it, for plain weaving. Shed rolls are made from

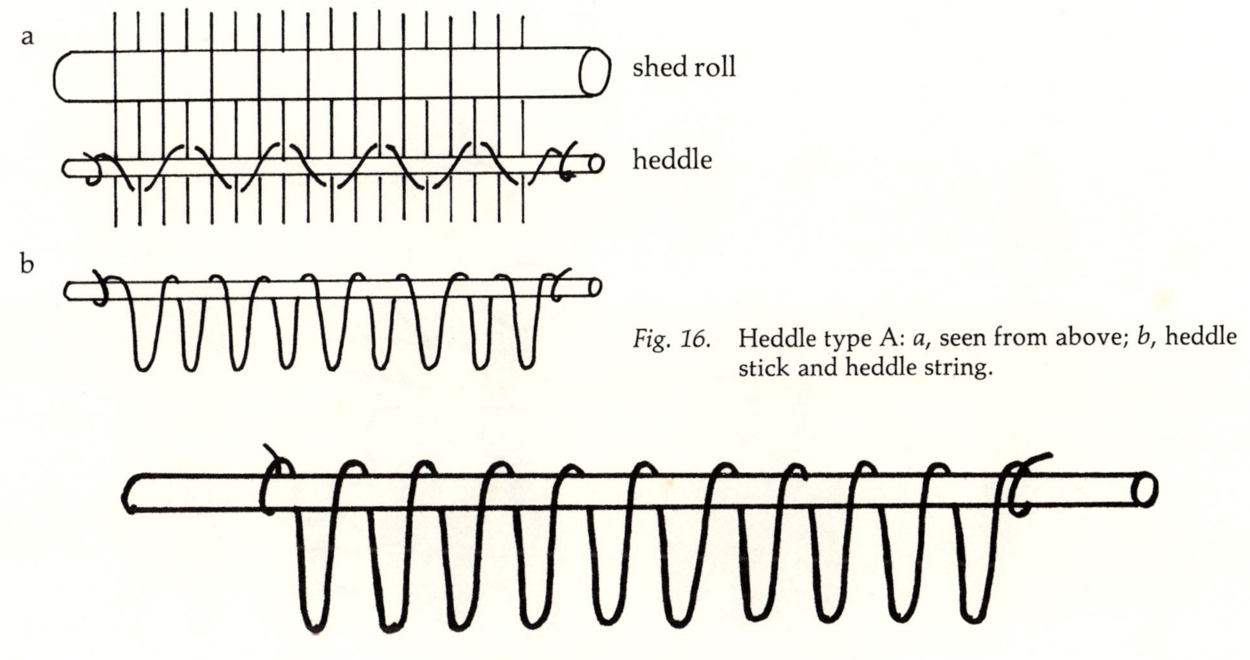

Fig. 16. Heddle type A: *a*, seen from above; *b*, heddle stick and heddle string.

Fig. 17. Heddle type B, warps omitted.

light wood such as pine, annona, or bamboo. They are somewhat longer than the end bars and often larger in diameter. In Nahualá, however, shed rolls and end bars are alike. In Aguacatán weavers use shed rolls made from bamboo. They put some seeds into the hollows and stopper them with corncobs. Whenever the weaver battens down the weft, the seeds make a rattling noise.

THE HEDDLE

The heddle controls the warps that pass under the shed roll. It consists of a strong and slender rod—the heddle stick—and the heddle string that connects the warps to the heddle stick. By lifting up the heddle stick, the heddle-controlled warps are raised above the warps that pass over the shed roll. The resulting shed is the countershed or *shed 2.*

Every time the heddle is lifted, the lower-positioned heddle-controlled warps are pulled past and above the warps that are controlled by the shed roll. If the warps are tightly packed, there is a lot of friction, and the heddle string sometimes breaks. A strong thin yarn is best for the heddle string. White sewing yarn no. 20 is used when available. Strong nylon thread is durable, but not every weaver can afford it. Where textiles have a fairly low warp count, thicker heddle strings can be used. In Nahualá weavers twist together several plies of yarn to make a heddle string. It is sized together with the warp to make it smoother and stronger. In Aguacatán weavers apply grease to the heddle string to reduce friction.

We observed two types of heddles, differing in the way the heddle string is looped around the heddle stick. Heddle type A, as we shall call it, is used as the standard heddle in all the places that we visited. Heddle type B is used in some places for special purposes.

Heddle Type A

This is the standard heddle. The heddle loops that pass under every second warp hang down in front or in back of the heddle stick in alternation. This

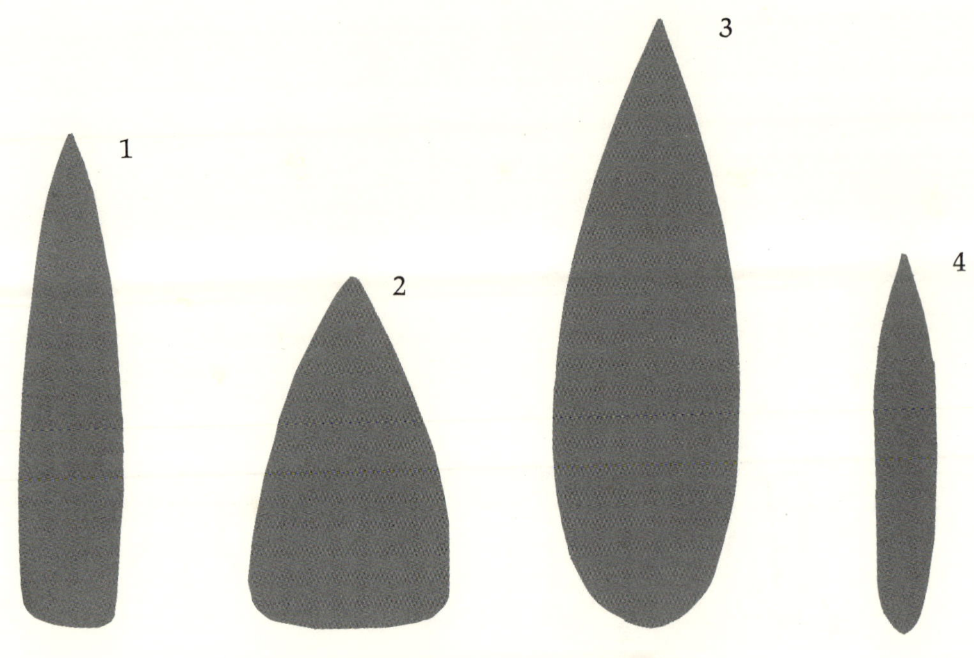

Fig. 18. Cross sections of four battens (actual size).

arrangement reduces the friction between the heddle string and the warps that pass over the shed roll (figure 16).

Heddle Type B

In San Sebastián Huehuetenango many weavers make a new heddle when weaving the join. Looping the heddle string around the heddle stick is difficult, because the warps are now so short (plate 33). The weaver then uses an easier method of looping the heddle string around the stick. The resulting heddle is shown in figure 17. This kind of heddle is also employed for weaving with wool.

THE BATTEN

Battens are used to help separate the odd- from the even-numbered warps, to beat down the warp crosses, to open a shed through which the bobbin can pass, and to batten down the weft. They have pointed ends to facilitate insertion between two layers of warps, and a straight edge to beat down the weft. To open a shed for the weft pick, battens are set on edge. In cross section some battens appear wedge-shaped, others are almost elliptical, many are shaped like a boat (figure 18). Battens and shed rolls in one loom setup have approximately the same length. The longest battens we saw were 46 inches (from Aguacatán) and 45 inches (Nebaj). Most weavers own several battens of different sizes.

Battens have a smooth surface and are rubbed with a little grease or oil to keep the friction between the wood and the warps to a minimum. Old battens have indentations on their working edges caused by the constant rubbing of the warps against the wood.

Weavers in Todos Santos, Aguacatán, Nebaj, Chajul, and Cotzal use heavy hardwood battens. In other places, where hardwood is not available, lighter woods, such as cyprus or even pine, are used. Occasionally battens are sold on the market.

Men from San Juan Atitán sell large battens for about sixty centavos. Hardwood battens are more expensive. In Chichicastenango and Chiché carpenters offer battens made from very light wood (probably pine) for sale. In San Martín Sacatepéquez many battens appear to be homemade. Elsewhere battens are made by specialists.

In figure 18 we show life-size cross sections of four battens. Batten 1 is 37 inches long and 2½ inches wide. It was used for the weaving of huipils and *tzutes*. The batten was made in San Juan Atitán and sold to a weaver in San Sebastián Huehuetenango.

Batten 2 was used for weaving belts. It is 16 inches long and 1¾ inches wide and fairly heavy. The batten was made in Todos Santos and sold to a weaver in San Sebastián Huehuetenango.

Batten 3 measures 18 inches long and 3 inches wide and comes from Aguacatán. It is made from hardwood and used for weaving belts.

Batten 4 is 36 inches long and 2 inches wide and comes from San Martín Sacatepéquez. It was used for weaving huipils. It is made from light wood, probably pine. Weavers in San Martín do not batten down the wefts. Instead they press the batten against the weft with a wiggling motion.

THE BOBBIN

Bobbins are slender sticks, slightly longer than the width of the fabric on the loom. They carry the weft yarn through the shed. Most bobbins are simple sticks with no notches or other provision for securing the weft yarn. In Nahualá, however, we encountered pointed sticks with slits, and in San Antonio Aguas Calientes weavers use reed bobbins with forked ends. In Todos Santos weft yarn is often hand-spun, and the spindle is used instead of a bobbin.

Loading the Bobbin

Unless the bobbin has forked ends, the weft yarn is wound five to ten times around one end of the bobbin. Then the yarn goes to the other end of the bobbin to take a few turns, and so on.

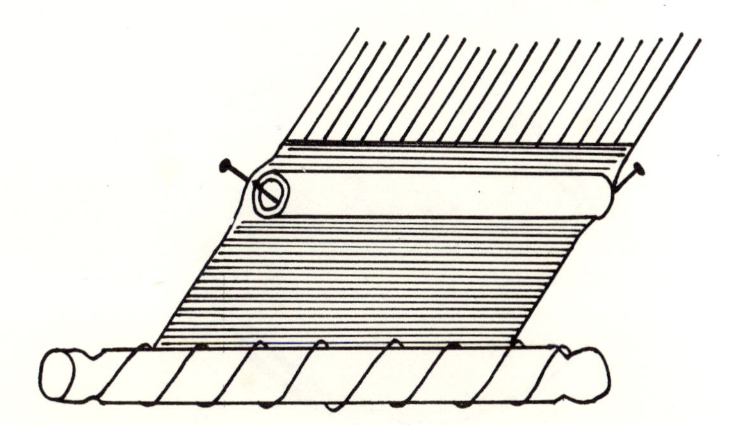

Fig. 19. Tenter, fastened to the underside of the weaving with two nails. The upper part of the loom and the backstrap are omitted.

THE TENTER

The tenter is a stick that controls the width of the fabric under construction. Tenters are used in all places that we visited. We encountered two kinds of tenters. Most widely used is a reed or bamboo stick that is cut so that its length is equal to the width of the weaving. The tenter is placed underneath the material, close to and parallel with the working edge, and is held in place by nails, needles, or spines that are stuck through the edges of the material and into the hollow of the reed.

In Todos Santos we saw tenters that can be used for fabrics of various widths. These tenters are made from cane and have a number of holes corresponding to the width of the material under construction.

THE CLOTH BAR

When the edge of the weaving has progressed to a point where the weaver cannot reach it easily,

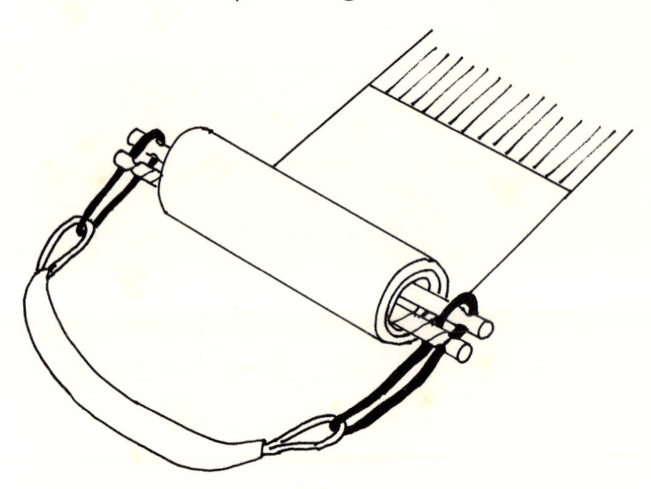

Fig. 20. Cloth bar, lower end bar, and backstrap. The woven material is wound around the two bars.

a cloth bar is placed parallel to the lower end bar, and the woven material is rolled around both the cloth bar and the end bar. Both bars together serve as cloth bars. Some authors refer to the extra bar as a supplementary cloth bar. Its main function is to keep the rolled up material from unrolling, as can be seen in figure 20.

Cloth bars are often of the same size and shape as end bars. Usually a set of three sticks is made at the same time, two of which serve as end bars and one as the cloth bar.

THE LEASE STICKS

Lease sticks are round or flat sticks. They are inserted into the warps behind the shed roll to establish warp crosses between the odd- and the even-numbered warps. While not absolutely necessary for the operation of the loom, lease sticks make weaving easier.

Lease sticks keep the warps in the right order at a point close to the shed roll. This is very convenient when weaving with long warps that tend to get entangled. Most important, the warps are kept in a fixed position by the lease sticks. This makes it easier to establish the countershed, as will be explained in chapter 9 under "Basic Weaving Procedures." Weavers also use the lease sticks to control the width of the material. When brocading is done with brocading wefts crossing the whole width of the material, there is a strong tendency for the warps to pull inward, and the tenter alone cannot keep the material at a constant width. It helps to spread out the warps in back of the shed roll (plate 43). The lease sticks keep the warps in this spread-out position, at least for a while.

Weavers in San Sebastián Huehuetenango, Santiago Chimaltenango, San Pedro Necta, San Rafael Petzal, Colotenango, and San Ildefonso Ixtahuacán do not use lease sticks. In other towns we saw looms equipped with one, two, or three lease sticks. Some weavers tie the lease sticks together with strings.

8

SETTING UP THE LOOM

Setting up a loom requires skill and patience. If it is not carefully done, the weaver will have to struggle with many problems during the weaving. While each weaver has her own way of setting up the loom, there are only two basically different setups in use. In the first kind the warps are tied to the end bars. This enables the weaver to weave to the very end of the warps. In the second kind of setup the warps pass around the end bars, and it is not possible to weave to the end of the warps. This kind of setup is used for weaving with end fringes.

LOOM FOR WEAVING WITH END SELVAGES

Most textiles are woven with the warps tied to the end bars. To set up the loom, the following steps are necessary: The warp chain as it comes from the warping board has two crosses (figure 21*a*). The weaver inserts two temporary end bars and a shed roll into the warps, as shown in figure 21*b*. She tensions the warps between the temporary end bars and arranges them in the right order on both temporary end bars. She then replaces the temporary end bars with end bars that are tied to the warps (figures 21*c* and 21*d*). A heddle is made at a point between the lower end bar and the first warp cross (figure 21*e*). The first weft pick is put into shed 2, the shed that is produced by the heddle. After weaving a heading strip, the weaver turns the loom around and starts weaving from the other end. Weaving continues until the edge of the weaving comes close to the heading strip. Then the weaver has to bridge the gap by making a join.

Weavers in different places have their own methods of doing the several steps for setting up a loom. In the following section we describe in detail the setting-up procedure as it was done by a weaver in San Sebastián Huehuetenango. In further sections we briefly discuss the setting-up procedures observed in other towns.

Setting Up the Loom for a Servilleta in San Sebastián Huehuetenango

Julia, the weaver, had warped the yarn for the

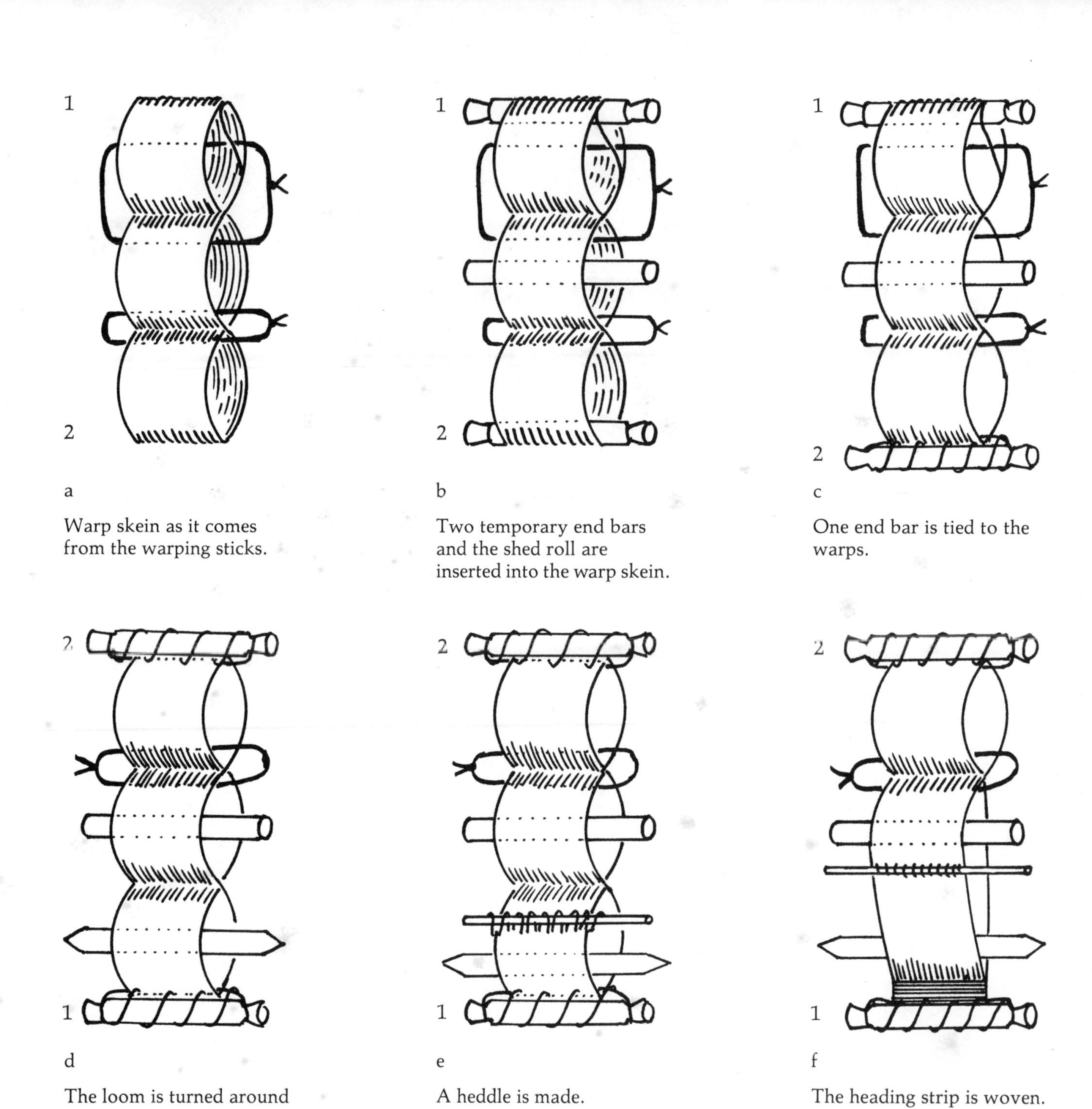

Fig. 21. Setting up a loom for weaving with end selvages, San Sebastián Huehuetenango. The backstrap and a rope hitching the loom to a house post are omitted in the diagrams, which are continued in figure 25.

Plate 17. Setting up the loom in San Sebastián Huehuetenango. The warps are spread out over the temporary end bars. Notice the strings securing the warp crosses.

servilleta on three sticks. For warping the stripes she used a *cadena* technique (crossing over of warps). After sizing the yarn, she selected loom sticks of appropriate length. The finished piece was supposed to be two *cuartos* (about 14 inches) wide. Julia used two of the warping sticks as temporary end bars. She washed the sticks and rubbed them dry. Before sizing, Julia had twisted the warp skein as if to turn it inside out. This was done to prevent tangling of the warp threads. Before placing the warps over the loom sticks, she restored them to the original state in which they had been when she took them off the warping sticks. She inserted two temporary end bars and a shed roll into the warps as shown in figure 21*b*. With a rope she hitched one of the temporary end bars to a house post. The other end bar was attached to the backstrap. The end of the warp skein with the knots and the *cadena* was placed on the upper end bar. Julia put the backstrap around her hips and sat down on her heels as though she were going to weave. Leaning back against the backstrap, she stretched the warps between the temporary end bars. She loosened the strings that were tied around the warp crosses (plate 17) and spread out the warps along the lower end bar. Since she could not reach the upper end bar, she called her sister to spread the warps out along the upper end bar. The warps were still wet from sizing. At this point, some weavers interrupt their work to let the tightly stretched warps dry. Julia, however, started to arrange the warps along the lower end bar.

Close to the lower warp cross, she separated the warps one by one and moved them over to the

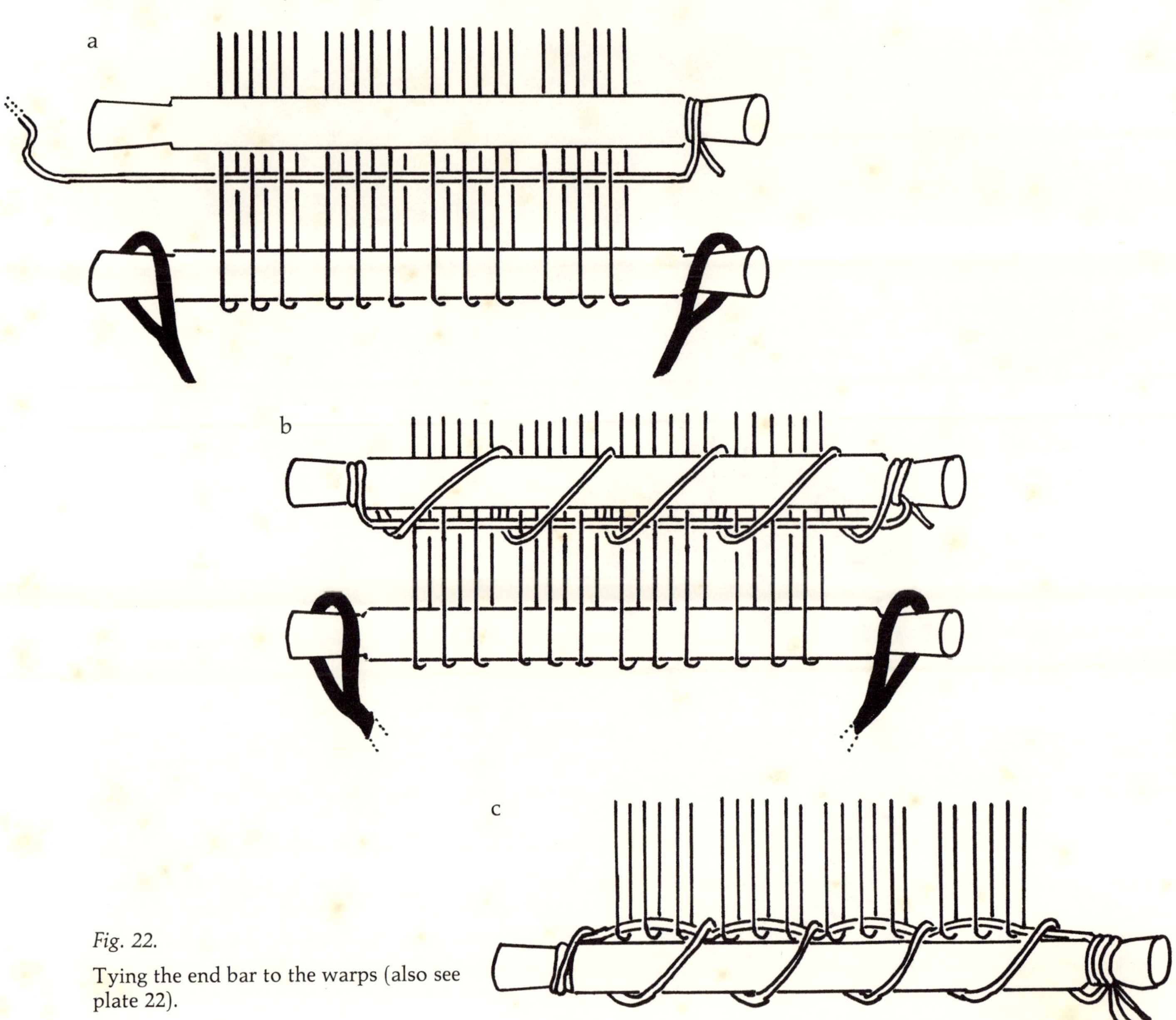

Fig. 22.

Tying the end bar to the warps (also see plate 22).

right. Some warps had been placed in the wrong position during warping, and Julia had to break them and tie them again. When all the warps were spread out in their proper order along the lower end bar, Julia arranged them in little groups slightly more than an inch wide, starting at the left side. She left small gaps between the groups. Later on, the loom string would pass through these gaps. Julia rearranged the warps once more so that they were spread out a total width of two *cuartos*.

Next she placed the stick that was going to be the permanent end bar on top of the warps, parallel to the temporary end bar. She tied a strong cord, made from maguey fiber, to the right end of the end bar. This cord—the loom string—was about three times as long as the end bar. Julia passed the loom string through the shed that was made by the temporary end bar (figure 22*a*). She pulled the loom string very tight and wound it several times around the left end of the end bar. Then she wound the loom string from left to right, looping it around the end bar so that it separated the groups of warps that were previously established (figure 22*b*). Julia tied the cord to the right end of the end bar and made sure that the string was tight and well fastened. She pulled the end bar downward and removed the temporary end bar from the loom (figure 22*c*). The end bar was attached to the backstrap, and once more Julia checked the position of the warps.

Next she turned the loom around and started to arrange the warps at the other end. This was the end where the knots and *cadena* (crossover of warp yarns) were made. Since the warps had already been put in the right order at the other end of the loom, Julia had no problem arranging the warps at this end (when working with striped material, it is easier to first arrange the warps at the end where there are no knots and warp crossovers). The way Julia had set up the loom, the warps that crossed over from one stripe to the next passed over the shed roll. Julia arranged the warps in the same order that they had been in on the warping sticks.

With all the warps correctly positioned, Julia tied the second end bar to the warps. She used the same method as for the first end bar. When placing the loom string into the shed made by the temporary end bar, Julia made sure that the string passed through the loops that connected the first and the last warps to the temporary end bar. She inserted a batten into the shed that was produced by the temporary end bar, withdrew the temporary end bar from the loom, and hitched the backstrap to the final end bar (figure 21*d*). The string that secured the lower warp cross was untied and retied so as to encircle the warps that passed over the shed roll. This was a precaution in the event that the shed roll were to slip out accidentally.

Now it was time to make the heddle. First Julia set the batten on edge. The warps that passed under the shed roll were on top of the batten. She placed

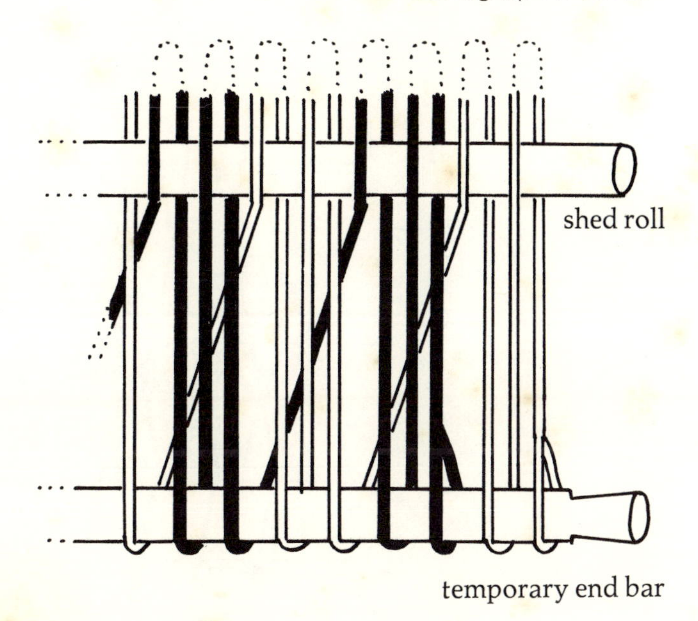

Fig. 23. *Cadena* (crossover of warps).

a small ball of white sewing yarn on the ground to her right. This was to be the heddle string. She inserted the yarn into the shed that was held open by the batten. At the left side, where the string reemerged from between the warps, Julia made a loop at the end of the string. She slipped this loop over the tip of the heddle stick, which she held in her left hand. Then she inserted her index finger between the first two warps at the left, close to the warp cross, and pulled up a loop of heddle string. This loop was placed around the heddle stick as shown in figure 24. Julia continued looping the heddle string in this way until all the warps passing over the batten were connected to the heddle stick (plate 18). She made the loops in such a way that one loop hung down in front of the heddle stick, the next loop hung down in back of the heddle stick, and so on. This is the standard heddle for weaving with cotton. While making the loops, Julia had to hold the heddle stick at the proper distance above the warps. The heddle loops are supposed to be a little shorter than the diameter of the shed roll. If the loops are longer, the heddle stick could slip out accidentally when the loom is tilted. When all the even-numbered warps had been connected

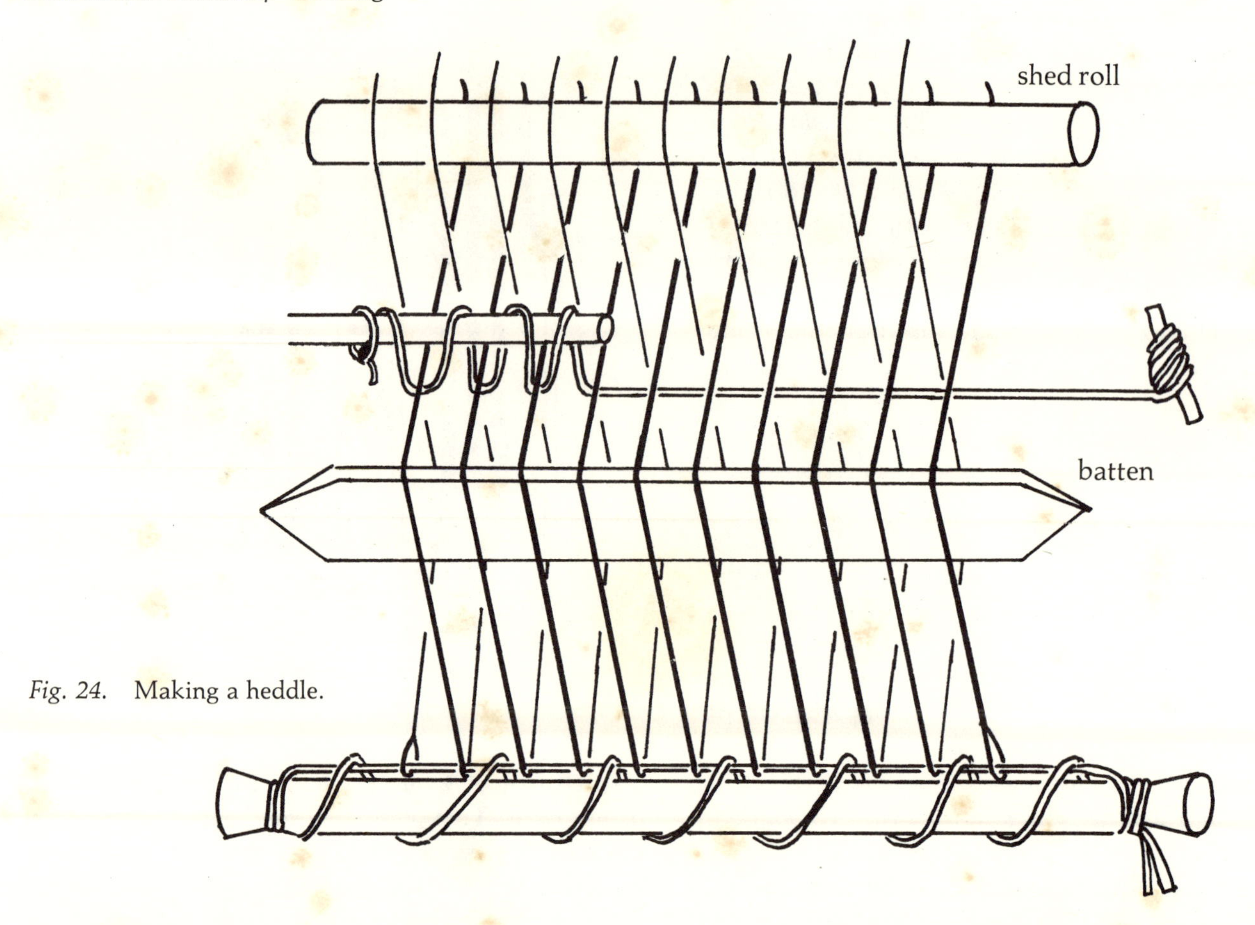

Fig. 24. Making a heddle.

to the heddle stick, the end of the heddle string was tied to the right side of the heddle stick.

With the heddle in place, Julia started to weave the heading strip. She had previously wound weft yarn on a bobbin. The first pick was put in shed 2 and went from right to left. Julia tied the end of the weft to the last warp on the right. After battening down the weft, she inserted the batten into shed 1 (the shed that is produced by the shed roll) and battened down the warp cross. Then she checked the warps once more for spacing and moved them into proper position, using her fingernails (plate 19). Gaps tend to form where the loom string passes between the warps, and Julia tried to keep these gaps as small as possible by rearranging the warps after each of the first few picks. She battened the wefts strongly to make a very firm heading strip. After 1½ inches of weaving, Julia made the last pick. It went from left to right and was put in shed 1.

Julia then pushed the shed roll and heddle toward the upper end of the loom and turned the loom around for the second time (figure 25*a*). She inserted an extra stick parallel with the batten. The batten was then withdrawn and reinserted next to the shed roll. The shed roll was put in the place that had been previously occupied by the batten. After batten and shed roll had exchanged places, the extra stick was taken out of the loom. The loom now appeared as shown in figure 25*b*. Next, Julia placed a weft into shed 2. To do this, she untied the string that secured the warp cross and used the string to pull the weft through shed 2 from right to left. Julia tied the end of the weft to the last warp on the left. After each of the next couple of weft picks she checked the warps for even spacing. It took her a long time to do the first inch of weaving. After she had woven two inches, Julia cut a reed that was as long as the width of the material under construction. She pinned this

Plate 18.
Making a heddle in San Sebastián Huehuetenango.

Plate 19.
Arranging warps in San Sebastián Huehuetenango. At the other end of the warps the heading strip has already been woven.

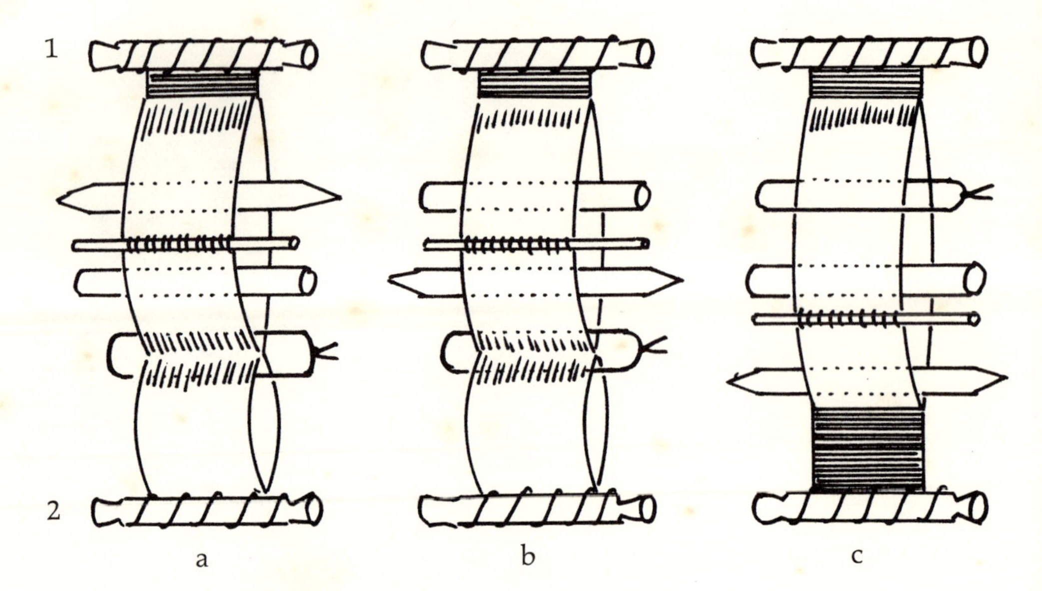

Fig. 25.

Rearrangement of loom sticks after the loom has been turned around for the second time, San Sebastián Huehuetenango (continuation of figure 21): *a*, the loom has been turned around for the second time; *b*, batten and shed roll have exchanged places; *c*, weaving begins at the lower end.

reed—the tenter—to the underside of the weaving with two nails. The loom was now set up, and no more loom sticks were necessary.

When setting up their looms, weavers in San Sebastián first tie the warps at *both* ends of the loom and *then* they make the heddle. This requires the loom to be turned around twice during the setting-up procedure. (After the warps have been tied at one end, the loom is turned around for the first time in order to tie the warps at the other end. The weaver then makes the heddle and weaves the heading strip. She turns the loom around one more time in order to start weaving at the second end.) We assume that the weaver sets up the loom in this order because she wants to establish the crossover of the warps (for the stripe pattern) before putting in the heddle. In the following sections we give brief outlines of setting-up procedures in other towns, where the weavers did not have to deal with crossover of warps. In all instances the heddle was made after or while the warps were arranged at the *first* end of the loom, and the loom was turned around only once during the setting-up procedure.

Setting Up the Loom for One Panel of a Huipil in San Juan Atitán

The weaver arranges the warps at the lower end of the loom and ties them to the end bar. She makes a heddle, weaves a very short heading strip and turns the loom around. Batten and shed roll exchange places. The weaver arranges the warps at the second side of the loom and ties them to the other end bar. She inserts two lease sticks into the warps.

Setting Up the Loom for One Panel of a Huipil in Chiché

The weaver arranges the warps at one end of the loom and then makes the heddle. Next she divides the warps into groups and ties the end bar to the warps. She weaves a heading strip and turns the loom around. Batten and shed roll exchange places. Warps at the other end of the loom are arranged and tied to the second end bar.

Plates 20 and 21. Arranging warps and making a heddle in Nahualá.

Setting Up the Loom in Nahualá, Chajul, and San Antonio Aguas Calientes

While arranging the warps at the lower end bar, the weaver makes the heddle *at the same time.* Starting at the left, warps are moved further to the left, and all warps that pass under the shed roll are connected to the heddle stick (plates 20 and 21). After the heddle is in place, the weaver checks the width of the warp setup and ties the lower end bar to the warps, as shown in plate 22. She weaves a heading strip. The loom is then turned around, and batten and shed roll exchange places. She arranges warps on the second temporary end bar and ties them to the permanent end bar. Two lease sticks are placed behind the shed roll.

LOOM FOR WEAVING WITH END FRINGES

Most belts and sashes are woven with end fringes, and such other textiles as shirts, *perrajes*, and blankets may have end fringes at least on one side. Plate 49 shows a loom that is set up for weaving with end fringes.

Setting Up the Loom in San Sebastián Huehuetenango

We observed the weaver in plate 49 setting up the loom to weave a woolen *delantera*. In the warp black and white wool alternate to form stripes. Warps of one color cross over the warps of the other color from one stripe to the next (*cadena*,

Plate 22.
Tying an end bar to the warps in Nahualá. The heddle is already in place.

as described in chapter 5). The weaver inserted end bars and shed roll into the warp skein as usual. The end where the *cadena* was made was placed at the upper end of the setup. The weaver arranged the warps at the lower end bar, turned the loom around, and arranged the warps at the other end. The string around the lower warp cross was replaced by two picks of wool weft. The weaver turned the loom around once more and made a heddle of type B (see chapter 7 under "The Heddle"). With the heddle in place, weaving was started at a distance of two inches from the lower end bar.

9

»»»»»»»»»»»»»»»»»»»»»»»»»»»«««««««««««««««««««««««««

WEAVING PROCEDURES FOR PLAIN WEAVING

The backstrap loom in its basic form is suited to plain weaving and variations of plain weaving (the basket weave, for instance). The two sheds necessary for plain weaving—we will call them shed 1 and shed 2—are established by manipulating shed roll and/or heddle. Shed 1 is kept open by the shed roll. Warps that pass over the shed roll will be called odd-numbered warps for the sake of convenience. Shed 2 is established by lifting the even-numbered warps above the odd-numbered warps with the aid of the heddle.

DIAGRAMS SHOWING WEAVING STEPS

To indicate positions of the shed roll, heddle, and batten in relation to the warps during weaving procedures, we employ diagrams showing the loom in side view. In the diagrams the two sets of warps —shed roll- and heddle-controlled warps—appear as lines. Only cross sections of the loom sticks are shown. The batten appears as a triangle, the shed roll as a black circle. The heddle is represented by a small circle (cross section of the heddle stick) and a loop symbolizing the heddle string. The end bars are omitted. The lower edge of the loom (where the weaver is working) is at the left side. For the sake of convenience, the loom is shown in horizontal position even though during the actual weaving it is always slanted. Figures 26*a* and 26*b* show a loom with the batten inserted into shed 1. Figure 26*c* shows a loom with the batten in shed 2.

WEAVING POSITIONS

Most women weave kneeling or sitting on their heels. This is convenient when supplementary wefts have to be added to the ground fabric, since utensils such as brocading swords and different colored yarns or pattern samples can be easily picked up from the ground. Many women stand and kneel (or sit on their heels) in turns. We saw a few women sitting on mats with their legs stretched out. In Chajul and Cotzal quite a few weavers sit on low chairs, probably to avoid sitting on the clay ground, which is always damp and cold in the rainy season.

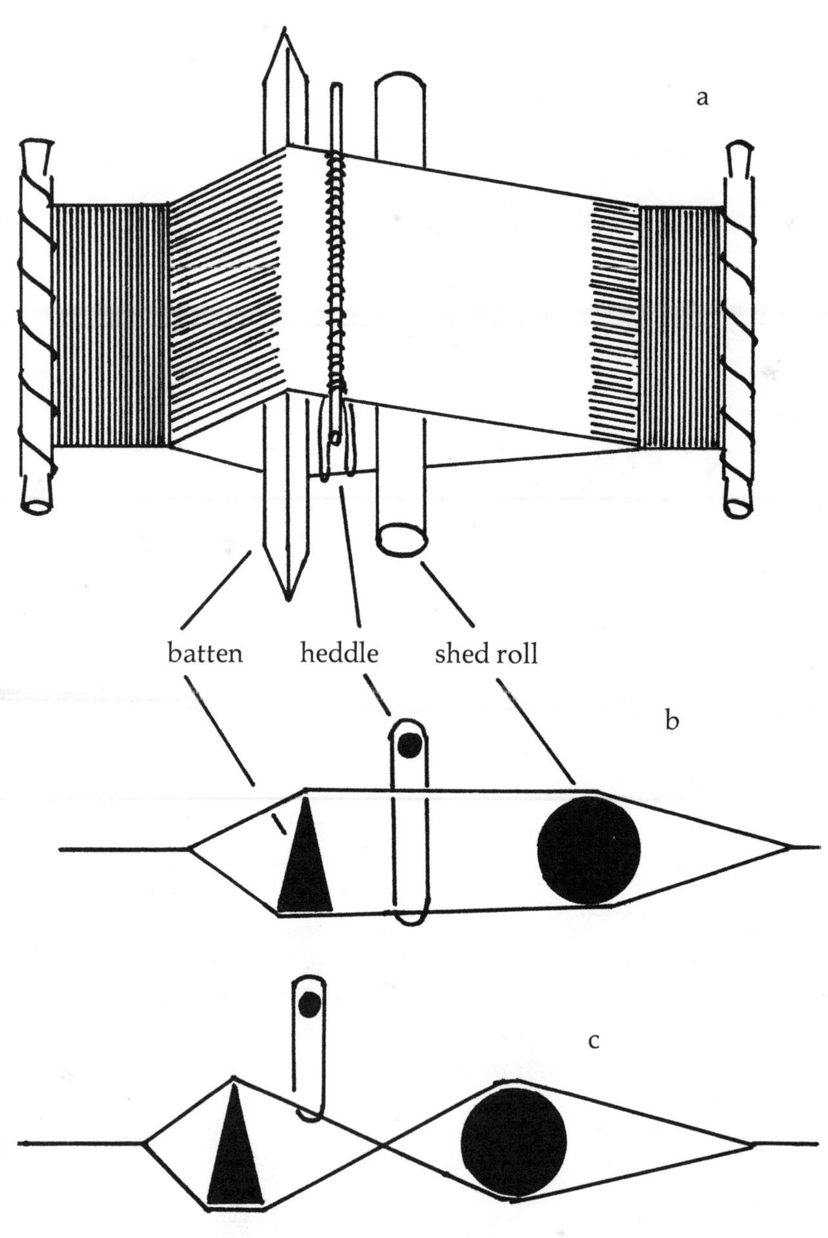

Fig. 26. Diagrams showing weaving steps: *a*, batten positioned in shed 1; *b*, batten in shed 1, side view; *c*, batten positioned in shed 2, side view.

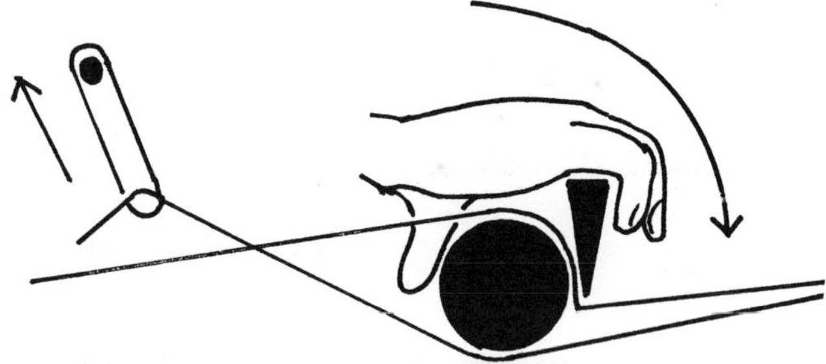

Fig. 27. Making shed 2, San Sebastián Huehuetenango

BASIC WEAVING PROCEDURES

The basic steps of backstrap weaving are:

1. Establish shed 2
2. Put a weft into shed 2
3. Batten down the weft
4. Establish shed 1
5. Put a weft into shed 1
6. Batten down the weft. Go to step 1.

To establish a new shed, the weaver has to move one set of warps past the other set of warps. When warps are tightly packed together, manipulation of shed roll and/or heddle alone is not enough to separate the two sets of warps. The weaver has to use the batten to clear the shed, and sometimes special procedures are required to make the warps separate. In the following section, we give a detailed description of the weaving procedures followed by a weaver in San Sebastián Huehuetenango.

WEAVING IN SAN SEBASTIAN HUEHUETENANGO

We observed a weaver working on a skirt. She preferred to stand while weaving.

1. In order to establish shed 2, Matilde, the weaver, moved heddle and shed roll away from each

Plates 23 and 24. While lifting up the heddle, the weaver presses the batten against the shed roll, moving both of them away from the heddle.

other. She leaned forward to slacken the warps. With her left hand she lifted the heddle. The batten was held in her right hand. She grasped the shed roll with her right hand while forcefully pressing down with the batten on the odd-numbered warps against the back of the shed roll (plate 23) until the batten was practically pushed under the shed roll; at the same time she pulled firmly on the heddle with her left hand (plate 24 and figure 27). This action separated the two sets of warps. Lifting the heddle alone would not have accomplished this, since the tightly packed warps cling to each other. Matilde now put the batten in shed 2 (plate 25). She moved heddle and shed roll toward each

Plate 25. The batten is put in shed 2.

Plate 26. The shed roll is moved close to the heddle.

Plate 27. The warp cross is battened down.

Plate 28. The weaver puts a weft pick into shed 2.

Plate 29. She battens down the weft.

Plate 30. She inserts the batten into shed 1.

Plate 31. She battens down the warp cross.

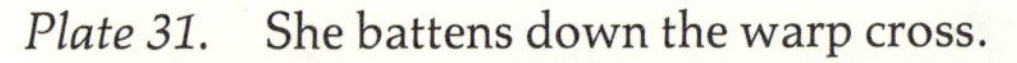

other (plate 26), then battened down the warp cross (plate 27) and set the batten on edge to open the shed for the weft pick.

2. She put a weft pick into shed 2 from left to right (plate 28).

3. The weft was battened down twice (plate 29) and the batten was withdrawn from the shed.

4. To make shed 1, Matilde moved shed roll and heddle close together and leaned back to put the warps under tension. She beat the warps in front of the shed roll with the back of her right hand, causing the odd-numbered warps to separate from the even-numbered ones. The batten was inserted into shed 1 (plate 30), and the warp cross was battened down (plate 31). Then she set the batten on edge to open shed 1 for the weft pick.

5. She put a weft pick into shed 1 from right to left (plate 32).

6. She battened down the weft twice and withdrew the batten from the shed.

When battening down a weft or warp cross, Matilde leaned back to tension the warps. Battening was done forcefully, as can be seen in plate 27. Heavy battening results in tight, durable material. When interrupting her work, Matilde stopped weaving after step 6. At this point the heddle stick and shed roll are close together and there is less danger of their slipping out. She disconnected the end bars from the rope and backstrap and rolled the woven material and bare warps over the lower end bar and the cloth bar. When she started weaving again, she began with step 1.

The weaving procedure as described above is fairly typical for weavers from San Sebastián. Weavers in other places often use different methods for making shed 2. Some of these methods are described in the next section.

Plate 32. She places a weft into shed 1.

DIFFERENT WAYS OF MAKING SHED 2

In Nebaj and Cotzal ceremonial huipils have a warp count of about 15 warps per inch. The warps have ample space between them and do not rub against each other. To establish shed 2, it suffices to lift the heddle.

With a higher warp count, it becomes more difficult to pull the even-numbered warps above the odd-numbered ones. The warps tend to cling to each other, and the heddle string causes additional friction. Weavers have developed various techniques for dealing with this problem.

In the foregoing section, we described one way of establishing shed 2, which we observed in San Sebastián Huehuetenango. In the same town, another method is also used:

The weaver lifts up the heddle with her left hand, and with her right hand she grasps the shed roll at the right side, pressing the odd-numbered warps against the shed roll and pushing the shed roll forward (figure 28). This separates the even-numbered warps at the right hand side from the odd-numbered warps. The weaver then moves her right hand a little to the left to press another section of warps against the shed roll, and again the shed roll is pushed forward. This procedure is repeated several times until all the even-numbered warps have been lifted above the odd-numbered ones. Weavers resort to this somewhat slow method when the warps are difficult to separate, especially when weaving with wool. Some weavers in San Sebastián Huehuetenango, especially beginners, use this method all the time.

In Chajul another method of establishing shed 2 is used. Weavers there employ very heavy hardwood battens that they leave in shed 1 when the heddle is lifted. With her right hand the weaver

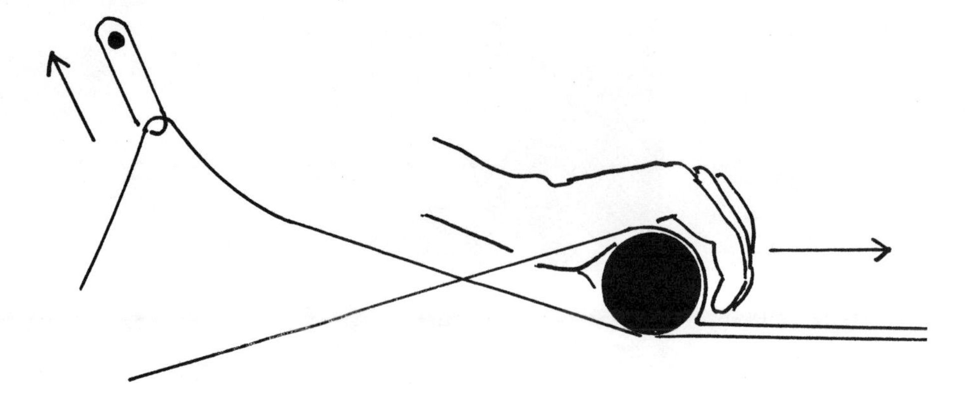

Fig. 28.
Making shed 2,
San Sebastián Huehuetenango.

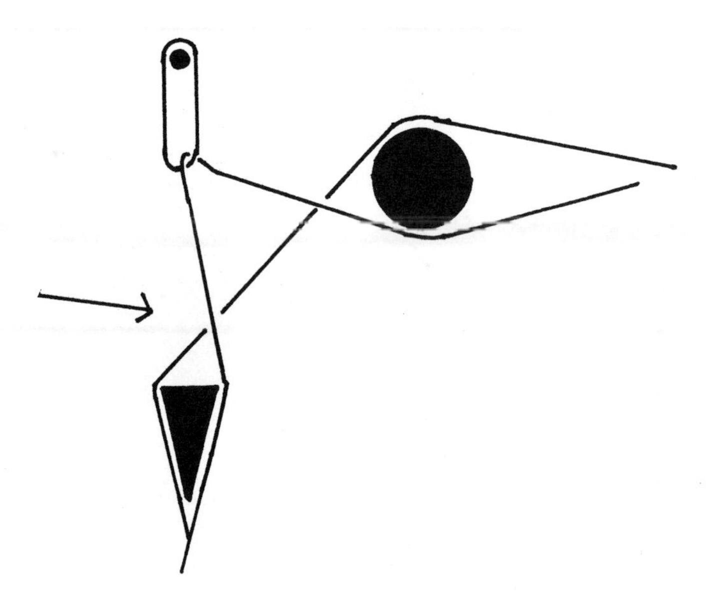

Fig. 29. Making shed 2, Chajul. The arrow indicates where the weaver hits the warp with the back of her hand.

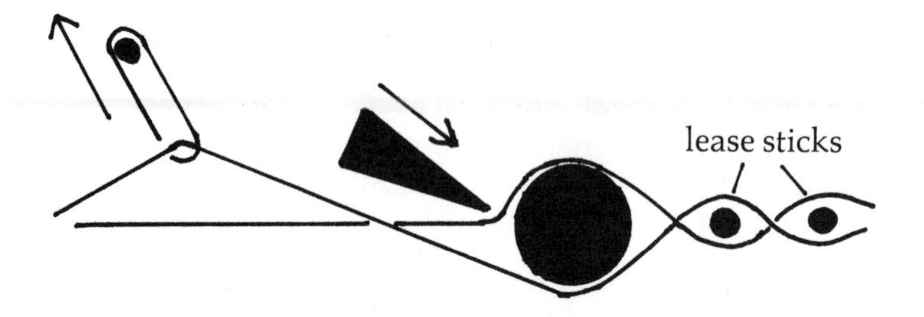

Fig. 30. Making shed 2, San Juan Atitán.

beats the warps above the batten (plate 44, figure 29). The warps separate, and the batten is withdrawn and put into shed 2. This method is rather tedious, but well suited to weaving material with a high warp count.

In San Juan Atitán (and other places where lease rods are used) we observed the following method: While lifting the heddle with her left hand, the weaver pushes the edge of the batten with her right hand against the shed roll until the shed roll touches the first lease stick (figure 30).

A variation of this method consists of pushing the shed roll against the lease sticks with the right hand while lifting the heddle with the left hand.

We found it easier to make shed 2 when the loom was equipped with lease sticks. Weavers in several *municipios* of Huehuetenango seem to think differently on this matter. They weave without lease sticks even though they must have seen lease sticks in nearby towns.

HEADING STRIP AND END SELVAGES

When weaving with end selvages, a heading strip is made at one end of the loom, then the loom is turned around and weaving starts at the other end. The heading strip is now at the upper end of the loom. Its purpose is to keep the warps in place and to enable the weaver to make a join. Some weavers

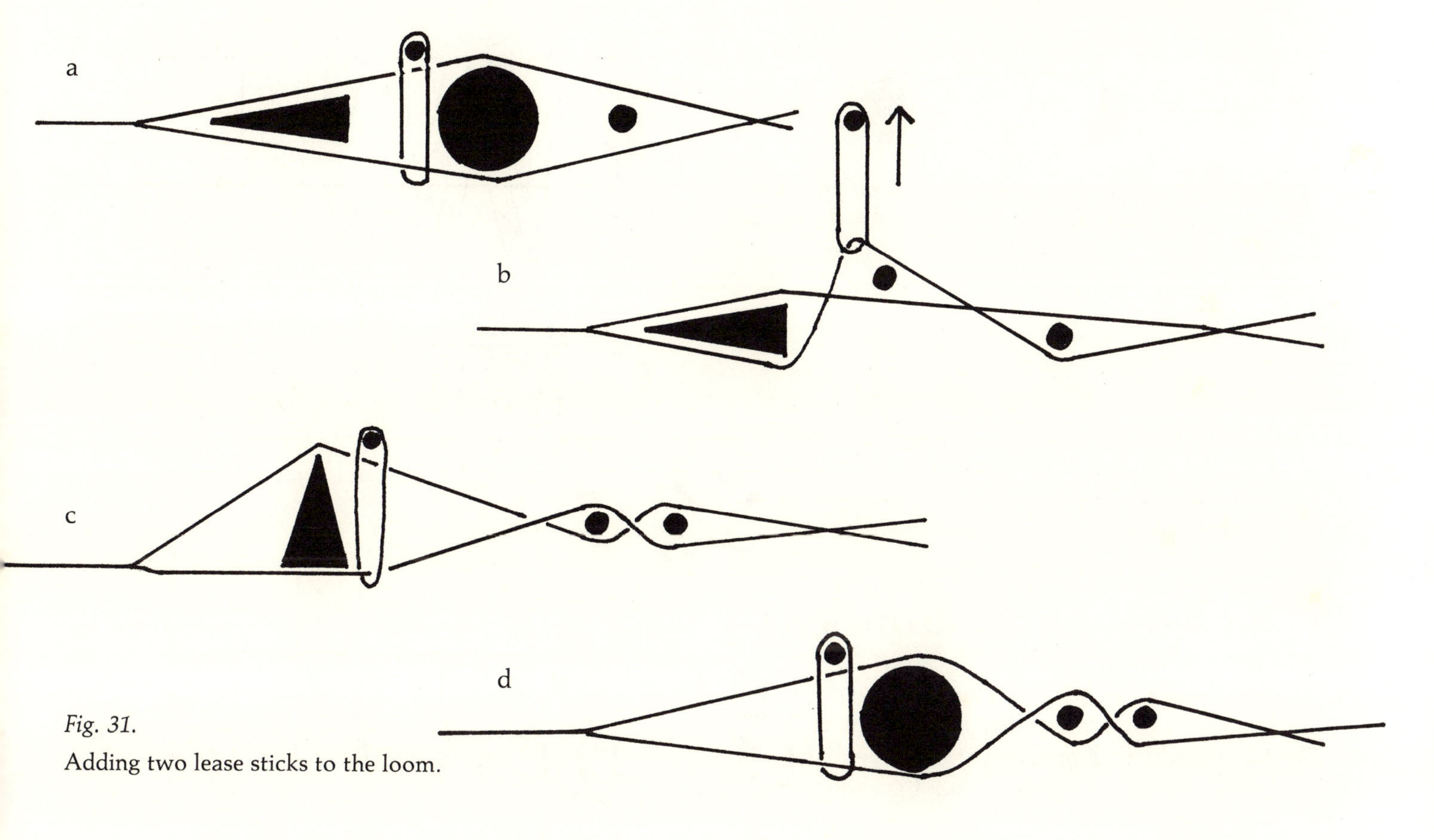

Fig. 31.
Adding two lease sticks to the loom.

make very short heading strips, consisting of a few picks. Heading strips 1 or 2 inches long are most common. In Nahualá heading strips are often longer than 10 inches.

To make the end selvages stronger, weavers double the weft for the first few picks at both ends of the weaving. Some huipils from Colotenango show wefts of double thickness for the first two inches or so. In San Martín Sacatepéquez and Almolonga the first (and sometimes the second) weft is many times thicker than the regular weft. The first few wefts at both ends of the loom are battened extra tight.

THE LEASE STICKS

Weavers in many towns equip their looms with one, two, or three lease sticks. In San Juan Atitán, we observed the following procedure for adding two lease sticks to the loom:

The weaver placed the batten in shed 1 and positioned the first lease stick directly behind the shed roll (figure 31*a*). She removed the shed roll and made shed 2 by lifting up the heddle. The batten was still in shed 1. The weaver put the second lease stick in shed 2, as shown in figure 31*b*. She set the batten on edge (figure 31*c*), moved the heddle toward the batten, and inserted the shed roll in shed 1 (figure 31*d*).

THE TENTER

After weaving a few inches the weaver fastens a tenter to the underside of the weaving. Often the

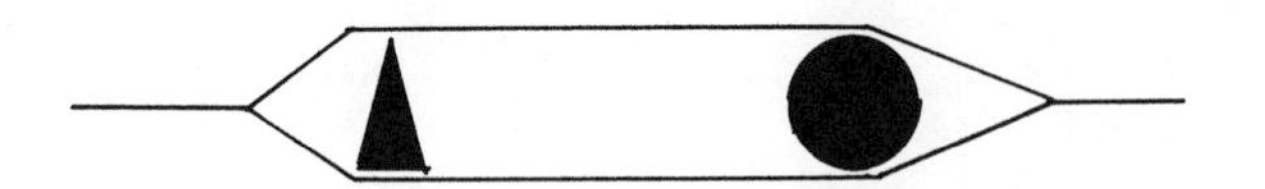

Fig. 32. Position of batten and shed roll prior to putting in a new heddle for weaving the join.

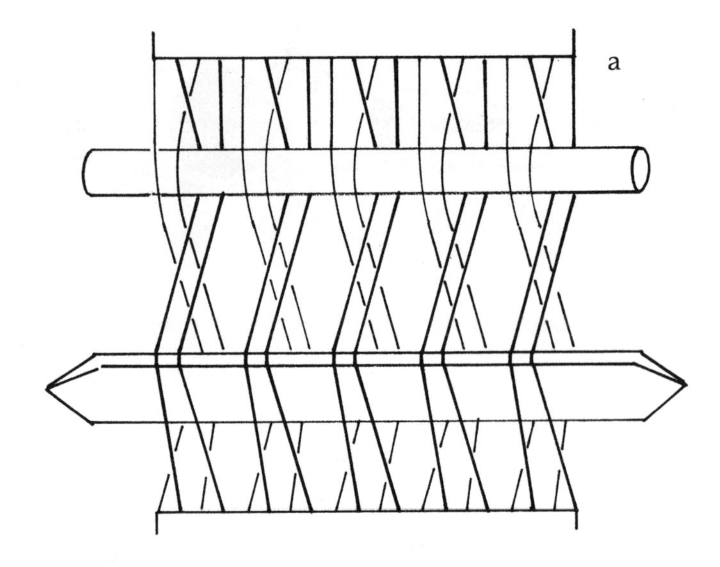

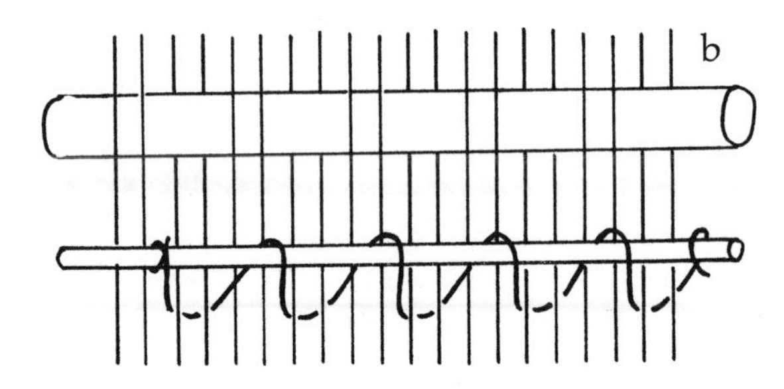

Fig. 33. The making of the heddle: *a*, warps are arranged in pairs to make a new heddle for weaving the join, San Sebastián Huehuetenango; *b*, the heddle for weaving the join.

tenter is made on the spot by cutting a reed to size. Without the tenter the woven material would get narrower as the weaving proceeds, since the weft tends to pull the warps closer together.

THE SUPPLEMENTARY CLOTH BAR

As the edge of the weaving moves farther away from the weaver, weaving becomes more difficult. To remedy this problem, the woven material is rolled around the lower end bar and a supplementary cloth bar.

ADDING NEW YARN

It is not necessary to tie weft ends together. Instead they are left to overlap in one shed.

THE JOIN

As the edge of the weaving approaches the heading strip, there is less space to manipulate batten and heddle. Nevertheless, most weavers continue weaving to close the gap. As the available space for manipulating the shed roll, batten, and bobbin becomes smaller, the weaver exchanges these items for smaller ones. Some weavers also make a new heddle. All weavers use thicker wefts to fill the gap faster. When the edge of the weaving comes very close to the heading strip, weavers employ needles instead of bobbins, and they use a comb or a fork instead of a batten to push down the wefts. Finally they bridge the remaining gap by pushing the wefts of the heading strip down toward the edge of the weaving with a comb. Standards for making a join seem to differ widely. In Colotenango, for example, most weavers make excellent joins that are hardly noticeable. In Nahualá, on the other hand, we saw many poorly made joins.

In the following, we describe in detail how a weaver in San Sebastián made the join on a *servilleta*. Warping and setting up of the loom for this *servilleta* have already been described in chapter 5 under "Warping Stripes" and in chapter 8 under "Setting Up the Loom for a *Servilleta.*"

As the edge of the weaving approached the heading strip, Julia, the weaver, replaced the standard batten and shed roll with smaller ones. When only 4 inches was left to weave, she also made a new heddle. Figure 32 shows the position of the shed roll and batten after the old heddle had been removed from the loom. The new heddle differed from the old one in three respects:

1. The heddle loops were shorter to match the new shed roll, which had a smaller diameter.
2. Each heddle loop controlled two warps instead of one. Working with only half as many loops as on the standard heddle reduced friction between heddle loops and the odd-numbered warps.
3. The heddle was of type B, probably because this kind of heddle is easier to make when the warps are short.

Before Julia could start making the heddle, she had to arrange the warps in pairs. Sorting started at the right-hand side, between shed roll and batten. Julia placed the first two odd-numbered warps at the right under her index finger. The first two even-numbered warps went over her index finger. In this way, the odd- and even-numbered warps were arranged in pairs and placed either over or under her index finger. After the length of her finger was covered with warp yarn, Julia put a string through the shed that was made by her finger. When all the warps had been arranged in this way, Julia inserted the batten alongside the string. The string was removed, the batten was set on edge (see *a* in figure 33) and a weft was put into the shed.

Next Julia passed white sewing thread through the shed and tied it to the heddle stick at the left-hand end. The heddle string was pulled upward between the pairs of even-numbered warps and looped around the heddle stick in a spiral (heddle type B). Plate 33 shows the making of the heddle.

With the new heddle in place, weaving continued. Figure 34 shows the transition from weaving with single warps to weaving with paired warps.

Fig. 34. Join with paired warps, San Sebastián Huehuetenango.

Because there was little space for manipulating batten, heddle, and bobbin, weaving became more difficult (see plate 34).

After two inches of weaving was done, the bobbin no longer fit into the shed and was replaced by a long, narrow metal rod taken from an old umbrella. Instead of a batten Julia now used a wooden comb to push down the wefts and the warp crosses. When only one inch was left to weave, the shed roll was replaced by the metal umbrella rod, which up to now had served as the bobbin. Instead of a bobbin Julia used a large sewing needle to draw the weft—four plies instead of two plies—through the shed (plates 35, 36). She had to do the weft picks in sections, each section as long as the needle.

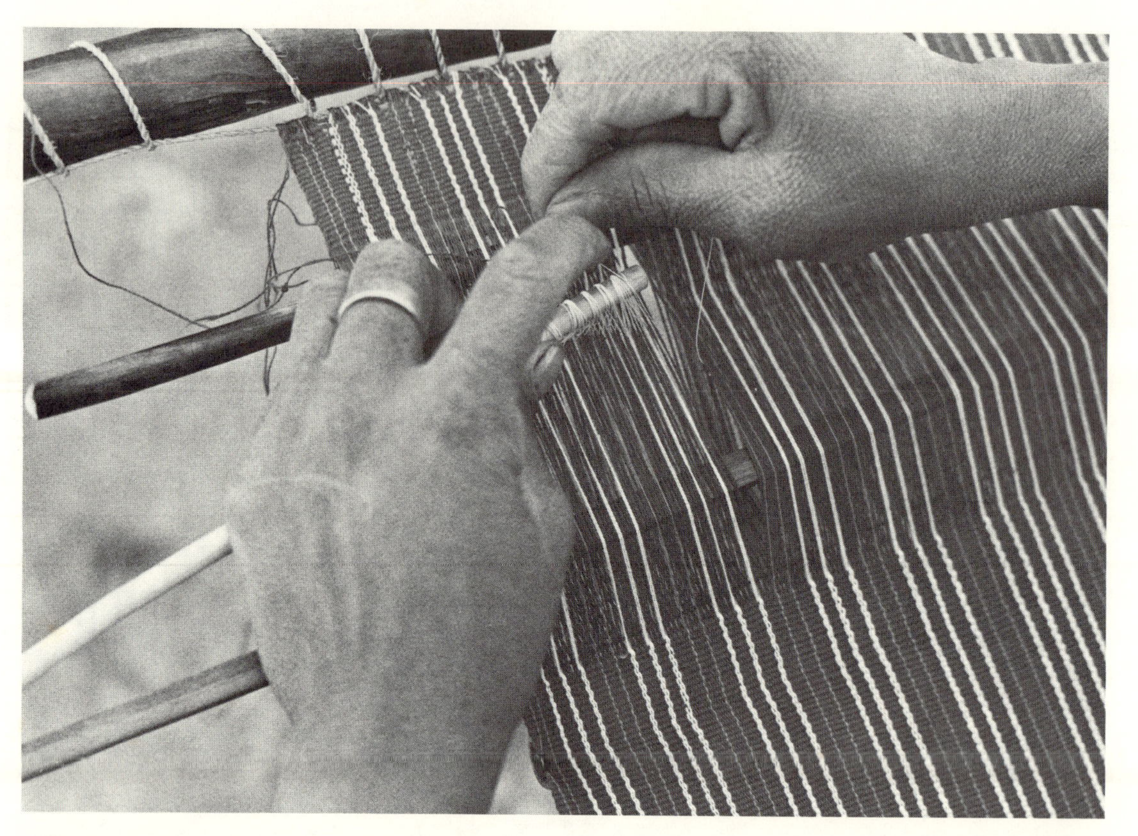

Plate 33. The weaver makes a new heddle. She is already using a smaller shed roll and batten.

Plate 34. The bobbin hardly fits into the shed.

Weaving the Join in San Sebastián Huehuetenango

Plate 35. The shed roll has been replaced by an umbrella rod. The weaver uses a needle to put the weft through the shed.

Plate 36. The wefts are pushed down with a wooden comb.

About eight weft picks were done with the needle. For the next-to-last pick, Julia tied the weft to the rod that served as the shed roll. She pulled the rod out of the loom, at the same time drawing the weft through shed 1. Next, Julia removed the heddle stick and started to wind the heddle string into a ball. The pulling out of the heddle string was done in stages. She wound the string that went into the making of the heddle loops, but she left the remaining string in shed 2. It served as a guide for the final weft pick, which ended at the left side. The final pick of the heading strip had also ended at this side. Julia tied both weft ends together. Next she pushed wefts from the heading strip downward with a comb. This was done to produce a smooth transition from the heading strip to the join. Finally Julia untied the loom strings that connected the finished piece of woven material to the end bar and took the material off the sticks.

10

»»»»»»»»»»»»»»«««««««««««««««

LOOM SETUPS AND TIME-SAVING PROCEDURES

When weaving one panel with end selvages or end fringes—the traditional loom setup—the finished piece can be taken off the loom without cutting any warps. This way no yarn is wasted. We assume that in the past, when weavers had to spin their yarn, it was standard procedure to set up the loom for only *one* panel. The time-saving practice of weaving *several* panels in one setup is probably of more recent origin.

Weaving with End Selvages

The warps are tied to both end bars. After weaving a heading strip at one end, the loom is turned around and weaving starts at the other end. Weaving is continued until the heading strip at the upper end is reached.

Weaving with End Fringes

The warps pass around both end bars. Weaving begins and ends a few inches from the end bars. A few weft picks are put in at one end of the setup. The loom is then turned around and weaving starts at the other end. Weaving is continued until the edge of the weaving comes within a few inches of the upper end bar.

Weaving with One End Selvage and Fringes at the Other End

Combining setups 1 and 2 produces a panel with a selvage at one end and fringes at the other. Warps are tied to one end bar and looped around the other one. A few weft picks are put in at one end. The loom is then turned around and weaving starts at the end where the warps are tied to the end bar. Weaving is continued until the edge of the weaving comes close to the upper end bar. It is not necessary to weave a join. After the material is taken off the end bars, the unwoven warp ends at one side are made into fringes.

Time-Saving Procedures

To save time, women in several *municipios* weave two (or more) panels in one setup. The warps are tied to the end bars as usual. Weaving is started at one end and is continued until nearly one half of the total length is woven. Then the weaver turns the loom around and starts weaving at the other end until she comes close to the weaving begun at the first side. It is not necessary to make a join. Instead the unwoven warps are cut to divide the material in half. Each half has a cut edge, which is then hemmed, and an end selvage. Had the two pieces been woven separately, more time would have been required to set up the loom twice and to make two joins.

Weaving with double- or multiple-length setups is common in Concepción Chiquirichapa and San Martín Sacatepéquez. In San Juan Atitán, *tzutes*, shirts, and pants are woven in double-length setups. The three panels that are needed for a huipil are woven separately.

In Todos Santos men's shirts are sewn together from seven pieces: two panels for the body, two panels for sleeves, two cuffs and a collar. One half of the shirt body and one sleeve are woven in one loom setup. Four cuffs and two collars are also woven together in one setup. The three panels for a huipil are woven separately.

In Almolonga and Zunil huipils are too long for a double-length setup. Weavers in these towns set up each panel separately, but they do not make joins.

Almolonga weavers tie the warps to one end bar and loop them around the second end bar. Weaving starts at the end where the warps are tied and continues until the edge of the weaving comes within one *cuarto* of the upper end bar. At this point, the unwoven warps are cut and the material is taken off the loom.

Zunil weavers tie the warp to both end bars and make a heading strip. The loom is then turned around and weaving starts at the other end. When the edge of the weaving comes close to the heading strip, the bare warps are cut and the material is taken off the lower end bar. Wasting a little bit of yarn seems to be acceptable to the weavers if it shortens the time for weaving.

Two-face brocading in San Juan Atitán.
The weaver uses a brocading sword to make a shed for the extra weft.

Spinning cotton in Todos Santos Cuchumatán.

Children from Todos Santos Cuchumatán.

Woman from San Martín Sacatepéquez.

Women and children from San Martín Sacatepéquez.

Market in Almolonga.

Procession in San Ildefonso Ixtahuacán.

Mother and daughter from Colotenango.

Above: Weaver from San Martín Sacatepéquez. The loom has three lease sticks that are tied together.

Below: Weavers from San Martín Sacatepéquez. Both women weave two panels in one setup. The woman at the left is weaving material for shirt sleeves. The woman at the right is weaving a huipil.

Above: Warping yarn for a huipil in Chiché.

Below: A Chiché weaver weaving a panel for a huipil.

Girl from Nebaj selling food at the market.

Girl from San Antonio Aguas Calientes.

Weaver from San Antonio Aguas Calientes making a belt. She uses the belt on her lap as pattern sample.

Young men from San Juan Atitán.

Men from Nahualá shearing sheep.

Women and children from Almolonga.

Market in San Rafael Petzal.
The women wearing white embroidered huipils come from Santa Barbara Huehuetenango.

Sales stand with yarn at the market in Colotenango.

Market in San Pedro Necta.

Market in Chajul.

Market in Zunil.

Procession in Zunil.

11

END FRINGES

End fringes are used as a finish for belts, sashes, blankets, *servilletas*, and sometimes also for shirts, *tzutes*, and *perrajes*. There are many ways of making end fringes.

TWISTED END FRINGES

Groups of warps are twisted together to form strands. We saw this kind of finish on a *servilleta* from Almolonga and on a shirt from San Juan Atitán. We also saw a similar kind of finish on a *tzute* from San Juan Atitán, with the addition of overcast stitches to prevent unraveling of the edge. The stitches were made while the material was still on the loom, just before it was cut through the center.

DOUBLE-TWISTED END FRINGES

Groups of warps are twisted together in an S direction. Two such strands are twisted together in a Z direction and tied at the end with a knot (figure 35*b*). This makes a very sturdy finish. We found it on sashes from San Juan Atitán and Zacualpa, shirts from San Martín Sacatepéquez, and *perrajes* from Aguacatán.

KNOTTED END FRINGES

Groups of warps are twisted into strands. The strands are knotted together as shown in figure 35*c* and plate 65. *Servilletas* from Concepción Chiquirichapa and Cotzal are often finished with knotted fringes.

BRAIDED END FRINGES

Women's belts from Santiago Chimaltenango have braided end fringes. Each braid consists of four strands, each strand is formed by several warps that are twisted together. The braids have knots at their ends. See figure 35*d*.

BRAIDED AND KNOTTED END FRINGES

Some *perrajes* from Cotzal are beautifully finished with braided and knotted end fringes. Groups of warps are twisted into strands. These strands are alternately knotted and braided. They end in braids to which tassels are tied.

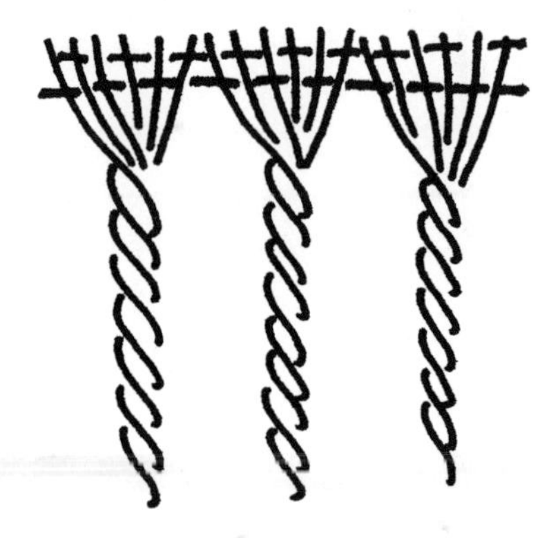

Fig. 35a. Twisted end fringes.

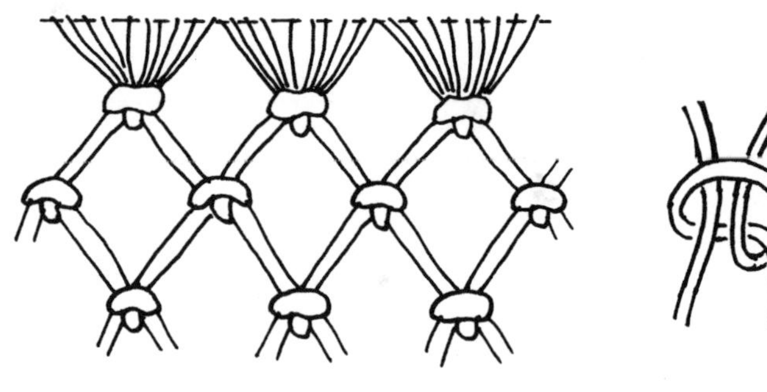

Fig. 35c. Knotted end fringes.

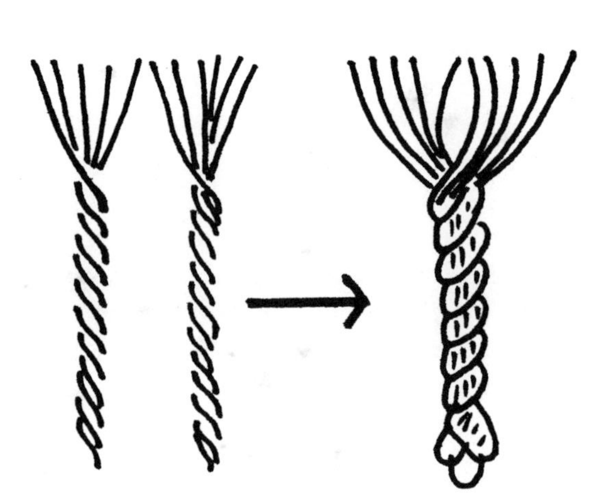

Fig. 35b. Double-twisted end fringes.

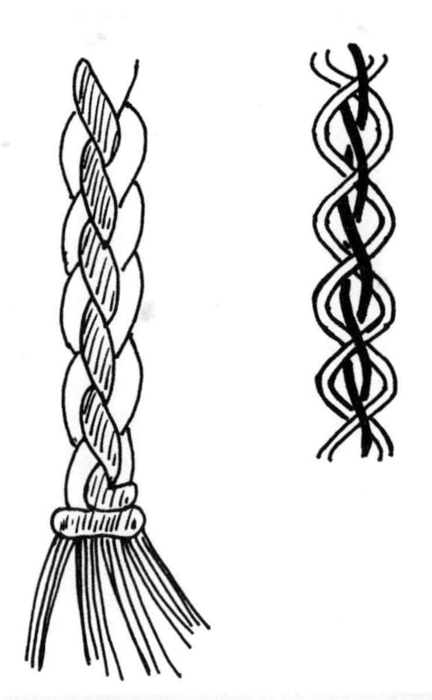

Fig. 35d. Braided end fringes.

12

»»»»»»»»»»«««««««««««

CONTRACTION OF THE WARP, OR TAKE-UP

Most fabrics that are woven on the backstrap loom have a warp surface (see chapter 16). Often, the warps are so close together as to hide the wefts completely. In such a fabric, the warps do not run in a straight line, but pass over and under the wefts as shown in figure 39. Because of this, the finished woven piece is considerably shorter than the straight warps as laid out on the warping board. A weaving with a relatively high warp count (60 and more warps per inch) and a weft count of 20 to 25 wefts per inch is up to 15 percent shorter than the warp before weaving. This effect is caused mostly by take-up. Weavers are aware of contraction and make the warps longer than the desired length of the finished piece.

Take-up is less noticeable in fabrics with a low warp count and negligible in fabrics with a weft surface. In the latter case the strong, widely spaced warps run straight, while the path of the thinner wefts curves over and under the warps. The wefts are placed into their sheds loosely to avoid a horizontal contraction or narrowing of the fabric.

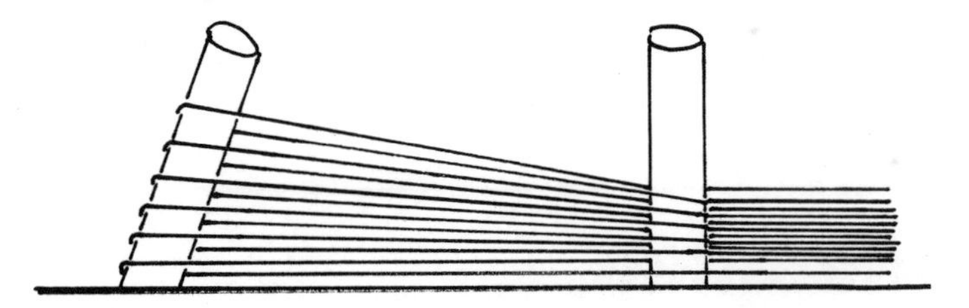

Fig. 36. Placement of warps on slanting warp peg.

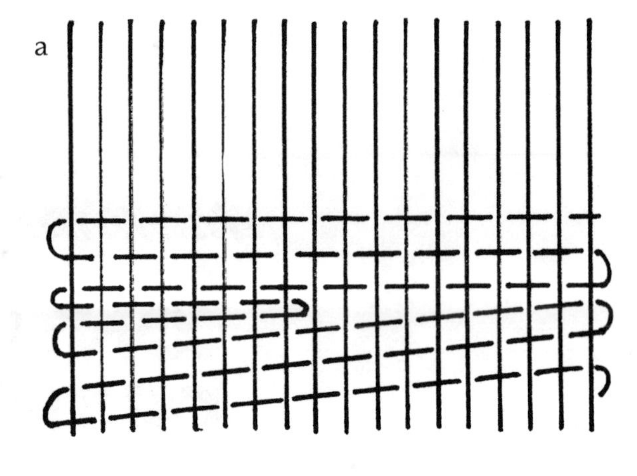

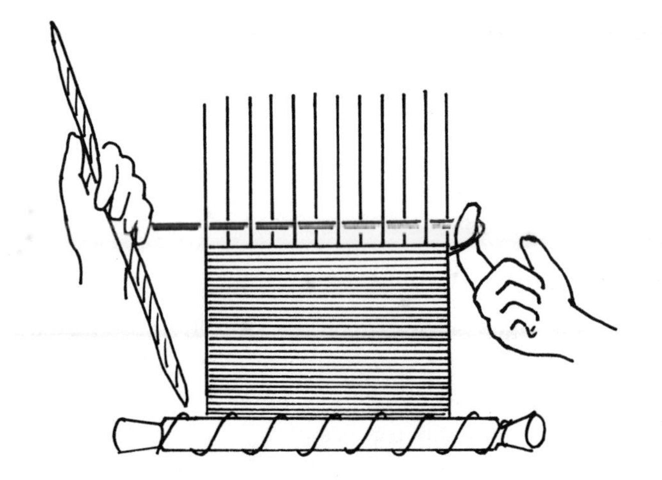

Fig. 37. Checking weft tension.

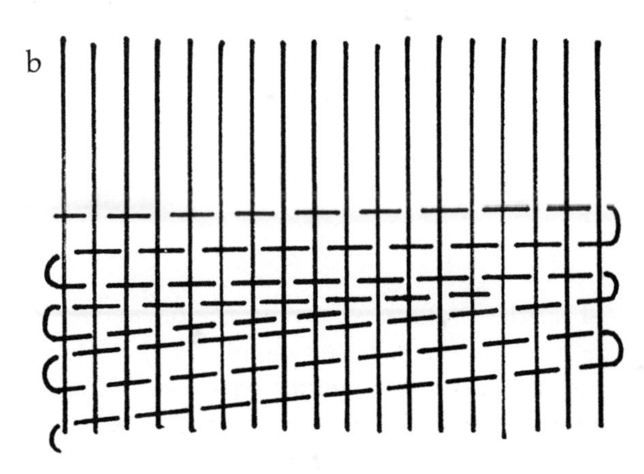

Fig. 38. Fill-in wefts.

13

TECHNICAL PROBLEMS, MISTAKES, AND THEIR CORRECTIONS

Skilled weavers make few mistakes and know how to correct them efficiently. Beginners and sloppy weavers frequently have to struggle with some of the problems listed below:

TWISTING

Single warp yarn that has not been twisted enough breaks easily during weaving. Yarn with too much twist tends to form kinks and is difficult to handle.

WARPING

Slanting warping pegs or sticks produce warps of different lengths. Often weavers try to minimize the effect of a slanting peg by putting the warps in a slanting position too (figure 36). Another method is to pile the warps on top of each other on the slanting peg or stick. Some weavers try to correct the effects of slanting warping sticks when setting up the loom. As long as the warps are still wet from sizing, they can be stretched. This is usually done by pulling with the fingers at one side of the warps when they are spread out on the temporary end bars. This method, like the other two mentioned above, does not work very well. In the end the weaver often has to resort to using fill-in wefts to solve the problem (see the section on "Fill-In Wefts" later on in this chapter).

During warping, warps are occasionally placed in the wrong position. Many weavers correct this mistake only when setting up the loom. Then the warps have to be broken and rearranged. To tie the broken warp, an additional piece of yarn is often necessary and two knots have to be made. Any knot in the warps is likely to cause problems unless it is made close to one of the end bars, out of the way of the heddle string, which tends to get caught in knots.

SIZING

Different yarns shrink different amounts during sizing. A weaver who uses white single yarn and colored two-ply yarn together in one setup has to cope with this problem. After the loom is set up, the colored warps are noticeably longer than the white ones. After weaving a few picks, the slack of the colored warps is transferred to the very edge of the weaving. The extra length is pulled out with a needle, forming a loop. If necessary, little sticks are placed under these loops.

SETTING UP THE LOOM

Uneven spacing of the warps and the loom strings that encircle the lower end bar results in gaps between warps close to the end selvages.

Some weavers are not careful enough in measuring the width of the piece; one end may be wider than the other.

WEAVING

Broken Warps

Broken warps have to be repaired with an extra length of yarn. If knots are made on heddle-controlled warps, they often interfere with the heddle string. This leads to further breaks in the warp or the heddle string. Some women opt for removing the broken warp altogether. Others take care to make the knots close to the edge of the weaving where they will not interfere with the heddle loops.

Weft Tension

The sides of the weaving need special attention. The weft should not be allowed to pull the warps inward at the sides. Weft that is put in too loosely, on the other hand, forms sloppy little loops or kinks at the sides. Special care has to be taken when the weft consists of several plies. The different plies should be under equal tension. Most weavers check the tension of the weft with their index fingers, as shown in figure 37.

Fill-In Wefts

If warping was done on slanting warping pegs, warps are not of uniform length and the end bars of the loom are not parallel with each other. Most weavers postpone dealing with this problem until they start making the join. Then they put in fill-in wefts at one side of the weaving. There are different ways of doing the fill-in. In figure 38, illustration *a* shows how the weft forms extra loops to build up the weaving at the left side. This method is fast and many weavers use it. The result, however, leaves something to be desired. A little hole results where the fill-in weft turns back because warps are pulled out of place. Illustration *b* in figure 38 shows a more elegant way of doing the fill-in, using an extra weft.

14

THE TIME FACTOR IN PLAIN WEAVING

Most weavers spend a few hours every day weaving. They can estimate how many days or weeks it would take them to finish a piece, but they could not compute the hours, since weaving is often interrupted by other occupations, and most women do not have watches anyway. When weaving for other people, weavers get paid by the piece.

We commissioned several women to work for us, and we timed the different procedures that are necessary to weave a piece of cloth. Since we paid by the hour, it is possible that some of the weavers worked slower than usual.

TWISTING—SAN SEBASTIAN HUEHUETENANGO

Many women in San Sebastián cannot afford to spend a few extra pennies on two-ply yarn. They buy the cheaper single-ply yarn and twist two singles together for the warp. We commissioned a woman to weave a panel for a *tzute* and computed the time she needed to prepare the warp yarn. The finished piece measured 23 inches × 40 inches. The preparation of six hanks of no. 16 single yarn took about nine hours.

Time Spent Preparing the Yarn

	Hours
Winding two singles together into balls	1½
Twisting the yarn with a spindle	6
Winding the twisted yarn into balls	1½
Total	9

In this instance, the preparation of the yarn (not counting the warping) took about as much time as the weaving!

WEAVING A SERVILLETA—SAN SEBASTIAN HUEHUETENANGO

The finished piece was 13½ inches wide and 25 inches long. On the warping sticks the warp was 29 inches (4 *cuartos*) long. The thread count is about 80 warps and 22 wefts per inch. Two-ply yarn (Mish) was used in warp and weft. The *ser-*

Plate 37. Weaving a strap for a bag in San Sebastián Huehuetenango.

villeta has thirteen groups of warp stripes, altogether sixty-five stripes.

Time Spent Making the Servilleta

	Hours
Winding the yarn into balls	1½
Warping (600 warp circuits)	2
Setting up the loom and weaving the heading strip	3
Regular weaving	4
Changing heddle for the join	1½
Weaving the join	2
Total	14

Of the fourteen hours it took to make the *servilleta,* less than half the time was spent on actual weaving.

WEAVING A PANEL FOR A HUIPIL—SAN SEBASTIAN HUEHUETENANGO

Huipils in San Sebastián are made from two panels that are woven separately. The finished piece measured 17 inches × 60 inches. On the warping sticks, the warp was 65 inches long. Two-ply Mish was used in warp and weft. The panel has about 150 warp stripes, 20 of which have at least two different colors (odd-numbered warps are of a different color than the even-numbered warps). The thread count is about 90 warps and 20 wefts per inch.

Plate 38. A girl from Nebaj weaving her first *servilleta.*

Time Spent Making a Panel for a Huipil

	Hours
Winding yarn into balls	3
Warping (765 warp circuits)	4½
Setting up the loom and weaving the heading strip	3½
Regular weaving	12
Join (heddle was not exchanged)	2
Total	25

Warping took a lot of time because of the many stripes. Most stripes were warped in the *cadena* technique. During the regular weaving (before reaching the join area), the weaver wove about a hundred picks or 5 inches per hour. Every so often she had to interrupt her work to look after her small children. When weaving without interruption, she made about two picks per minute, which is a little better than average.

WARPING AND SETTING UP THE LOOM FOR A SIDE PANEL OF A HUIPIL—CHICHE

The finished piece measured 12 inches × 64 inches. Two-ply Mish was used for weft and warp. Huipils in Chiché do not have warp stripes. The warp yarn (two-ply Mish) was wound from the reel directly onto the warping board. Warping (360 warp circuits) took a little less than two hours. Sizing the warps, setting up the loom, and weaving the heading strip took a little more than two hours. Altogether, four hours of work was required before the regular weaving could begin. Compared to the time it takes weavers in San Sebastián to get the weaving started, this is very fast. In the end, however, Chiché weavers have to spend much more time to finish the weaving, because a large part of the huipil is covered with pattern wefts.

These few examples show that the time needed for preparing and warping the yarn and setting up the loom depends on many factors: the kind of yarn used, the number of warps, the number of warp stripes, and of course the skill of the weaver. Weaving speed (picks per hour) depends on the skill of the weaver and the kind of cloth being woven and its width. A small loom is easier to handle than a large one, and material with a high thread count requires more effort than a loosely woven piece.

15

PLAIN WEAVE AND ITS VARIATIONS

The backstrap loom in its basic form is designed for plain weaving. Variations of plain weave are produced by combining several threads in warp and/or weft. In the following, we list some of the most common combinations of warp and weft elements.

Plain Weave (1:1)

This weave is characterized by one thread in warp and weft. It is used by weavers in Colotenango, San Ildefonso Ixtahuacán, San Sebastián Huehuetenango, and San Antonio Aguas Calientes. Yarn of the same size is used for warps and wefts. Weavers in San Antonio prefer strong three-ply yarn. In the other towns mentioned, two-ply yarn or hand-twisted singles (size no. 8 or no. 10) are used for the warp and singles (without extra twisting) are used for the wefts.

Paired Warps and Wefts (2:2)

This combination is found in huipils in San Juan Atitán, San Pedro Necta, Nahualá, in *cofradia* huipils in Chajul, and in skirts in the Huehuetenango area. Only single yarns are used. In San Juan Atitán, the singles are often hand-twisted and then paired for the warps. Threads of the same size are used for warps and wefts.

Tripled Warps and Wefts (3:3)

This weave is used in huipils in Concepción Chiquirichapa and San Martín Sacatepéquez. Thin yarn—no. 16 or no. 20—is used for warps and wefts.

Single Warp, Paired Wefts (1:2)

This variation is seen in huipils in Todos Santos, San Ildefonso Ixtahuacán, Zacualpa, and other places. Often the warp consists of two-ply yarn, the weft of paired singles.

One Thread in the Warp, Tripled Weft (1:3)

Weavers in Zunil use two-ply yarn for the warp and combine three singles of no. 8 yarn for the weft. Because of the strong weft, weaving builds up quickly.

Paired Warps, Tripled Wefts (2:3)

Huipils and skirts in Santiago Chimaltenango, shirts in Nahualá and San Martín Sacatepéquez, and *perrajes* in Cotzal are all made by this method.

Tripled Warps, Quadrupled Wefts (3:4)

This weave is found in use in huipils in Chajul and Nebaj. Warps and wefts are of the same size.

Three Threads in the Warp, Six in the Weft (3:6)

This combination is used in huipils in San Martín Sacatepéquez.

More variations of plain weaving are described in chapter 20.

16

»»»»»»»»»»»»»»»««««««««««««««««

THREAD COUNT, WARP SURFACE, WEFT SURFACE

Within one piece of fabric, the number of warps per inch is not uniform. It is highest close to the side selvages and considerably lower near the center. The wefts are more evenly distributed, except for the join area, where they cannot be tightly battened down.

In the chart that follows we list average thread counts (for plain weaving) from twenty-three textiles from our collection. The textiles are grouped according to the warp count. Where warps or wefts consist of several threads, they are counted as one. All items listed are woven with cotton yarn. In the chart the textiles are divided into three groups:

GROUP 1: FABRICS WITH A TRUE WARP SURFACE

The warps are packed together closely and cover the wefts completely. Textiles numbered from 1 to 15 on the list belong to this group. They have warp counts from 60 to 96 warps per inch and weft counts from 17 to 30 wefts per inch. The warp count is three to four times higher than the weft count. Plates 76 and 80 show textiles with a true warp surface. Textiles in group 1—the majority of the textiles listed—are very durable. Warp stripes show well on them, weft stripes are almost invisible.

GROUP 2: WARP SURFACE LESS PRONOUNCED

The warp count is more than two times and less than three times higher than the weft count. The wefts are visible between the warps. Textiles numbered from 16 to 22 belong in this group. The warp counts range from 42 to 56 warps per inch, weft counts go from 17 to 28 wefts per inch.

Thread Counts

Number	*Town*	*Article*	*Weave*	*Warps per Inch*	*Wefts per Inch*
1	San Rafael Petzal	Huipil	1:1	96	29
2	Colotenango	Huipil	1:1	92	30
3	Santiago Chimaltenango	Huipil	2:3	92	21
4	San Juan Atitán	Shirt	2:2	92	24
5	San Sebastián Huehuetenango	Huipil	1:1	90	23
6	Zacualpa	Huipil	1:1	90	20
7	Chajul	Huipil	3:4	82	21
8	Colotenango	Huipil	1:1	80	22
9	San Ildefonso Ixtahuacán	Huipil	2:2	76	21
10	Zunil	Huipil	1:3	74	17
11	Cotzal	*Perraje*	2:3	74	25
12	Todos Santos	Huipil	1:2	72	24
13	Nebaj	Huipil	3:4	70	18
14	San Martín Sacatepéquez	Shirt	2:3	60	23
15	San Antonio Aguas Calientes	Huipil	1:1	60	20
16	Concepción Chiquirichapa	Huipil	3:3	56	25
17	San Juan Atitán	Huipil	2:2	56	22
18	Nahualá	Shirt	2:3	54	28
19	Nahualá	Huipil	2:2	52	25
20	San Pedro Necta	Skirt	2:2	52	20
21	San Pedro Necta	Huipil	2:2	42	17
22	San Martín Sacatepéquez	Huipil	3:6	42	20
23	Nebaj	Huipil	6:3	14	36
	Stripes on same item		6:1	14	70

GROUP 3: FABRICS WITH A WEFT SURFACE

There is only one example on the list. Ceremonial huipils from Nebaj and Cotzal have weft surfaces. The warp count is very low—about 14 warps per inch—and the weft count varies in different sections of one piece. Textile no. 23 on the list, a huipil from Nebaj, has a weft count of 35 wefts per inch in most sections and 70 wefts per inch in the tightly battened weft stripes (plate 62). The time needed for weaving a fabric in plain weave is approximately proportional to the total number of weft picks, and textiles with a weft surface require more weft picks than textiles with a warp surface. This might be one reason why few textiles with a weft surface are woven on the backstrap loom.

FABRICS WITH A BALANCED THREAD COUNT

Missing on our list are textiles with a balanced thread count (equal number of warps and wefts per inch). Loosely woven textiles of this kind are made mainly in the department of Alta Verapaz. In the area of our study, they do not occur.

17

»»»»»»»»»»»»»»»»«««««««««««««««««

ELEMENTS OF DESIGN: OVERVIEW

Elements of design are incorporated into the fabric at different stages of fabric construction:

1. During the warping (warp stripes).
2. During the weaving (weft stripes, warp floats, different kinds of weave, extra wefts).
3. After the material has been taken from the loom (braided, twisted or knotted end fringes, sewing, embroidery, appliqué).

In chapter 20 textiles featuring different design elements will be described in detail. In chapters 17 and 18 we discuss the basic methods of incorporating designs into the weaving.

WARP STRIPES

Warp stripes can be found on many weavings that have a warp surface. Plain stripes are produced by odd- and even-numbered warps of the same color. A second kind of stripe is made when odd- and even-numbered warps are of different colors. Let us assume that the odd-numbered warps are white, the even-numbered warps black. With one weft pick the white warps come to the surface, with the next weft pick the black warps appear, and so on (figure 39).

In many places weavers have developed distinctive stripe patterns and color schemes. Almost all the textiles from the Huehuetenango area are striped, the most popular color combination being red and white.

WARP FLOATS

Warp floats are used to decorate women's belts. In Totonicapán and Huehuetenango belts with complex designs are made by specialists. In Santiago Chimaltenango, women's belts have simple warp float designs. Weavers in San Antonio Aguas Calientes make belts in the Totonicapán style and, using the same technique, weave belts that are decorated with birds, butterflies, and floral designs of European origin.

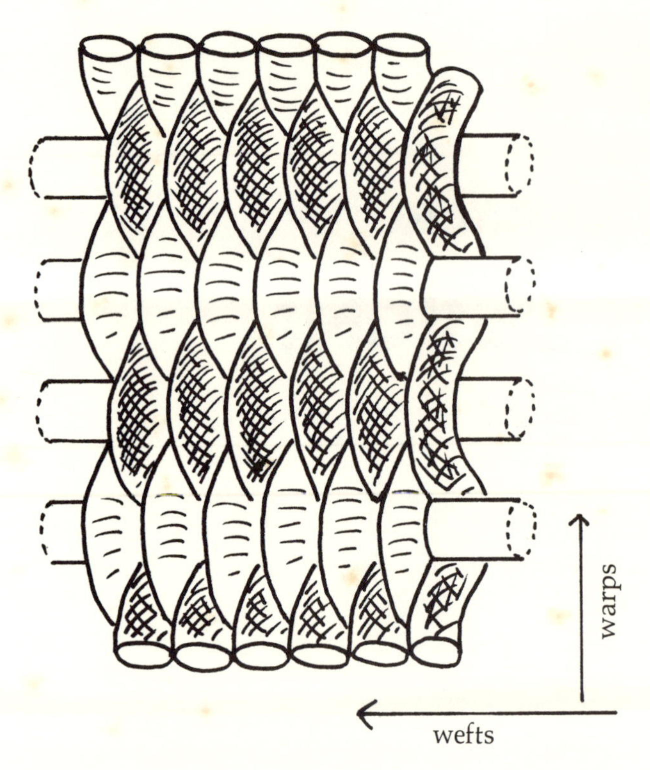

Fig. 39. Odd- and even-numbered warps of different colors.

WEFT STRIPES (PLAIN WEAVE)

Weft stripes are most effective in material with a weft surface. They can also be found on fabrics that have a fairly high warp count, as long as the wefts can be seen through the gaps between the warps.

We found textiles with weft stripes in Almolonga (huipils), Nebaj and Cotzal (huipils, *servilletas*), San Sebastián Huehuetenango *(delanteras)*, Todos Santos (men's sashes), and Zunil (old-style huipils).

WEAVES OTHER THAN PLAIN WEAVE

Weaves other than plain weave and its variations are employed to produce special effects. They are more time-consuming than plain weave and can be done only on looms with additional shed sticks and heddles. Weavers in Concepción Chiquirichapa and San Martín Sacatepéquez use twill weaves to decorate huipils with horizontal bands. In Todos Santos and San Juan Atitán, weavers create tapestrylike weft surfaces with a compound weave that requires two sets of wefts. Weavers in Santiago Chimaltenango have developed a special weave to produce weft stripes. Gauze weaving is done in the department of Alta Verapaz. In the geographical area of our study we did not encounter any gauze weaving.

TWINING

Textiles from Zacualpa feature bands that are done in a twining technique. In Colotenango and San Ildefonso Ixtahuacán twining is occasionally done with extra wefts.

EXTRA WEFTS

Weavers in most towns decorate huipils and other textiles with extra-weft patterns. San Sebastián Huehuetenango and Zunil were the only towns we visited where weavers made no use of extra wefts.

Extra wefts are added to the ground fabric between picks of the regular weft. Taking away the extra wefts would leave the ground material intact. *Continuous extra wefts* run across the whole width of the fabric to build up horizontal bands or rows of geometrical figures. *Discontinuous extra wefts* are employed for weaving free-standing figures.

Inlaid Extra-Weft Patterns

Inlaid extra wefts are placed into the same shed as the basic weft, wherever the design requires it. This technique is used extensively in the department of Alta Verapaz, where textiles are loosely woven and inlaid extra wefts show well. In the geographical area of our study most fabrics have a warp surface, and extra wefts are hardly visible between the tightly packed warps. Therefore if weavers put

Plate 39. Weaving material for a pair of pants in Todos Santos Cuchumatán. Since the weft yarn was handspun, the weaver uses the spindle as a bobbin.

extra wefts into sections of the basic sheds, it is done in order to hide the extra wefts from view between surface floats or to hide the beginning and end of an extra weft. Often the inlaid ends of extra wefts form cords or streamers, because the warps that pass over them are raised a little (plate 74).

Single-Face, Two-Face, and Double-Face Patterns

When the extra wefts are visible only on one side of the fabric, we speak of single-face patterns. The term two-face indicates that two different patterns appear on the two sides of the fabric. A fabric with identical patterns on both sides is double-faced. When making a single-face pattern, the weaver works with one half of the warps at a time (either the odd-numbered or the even-numbered warps). This type of pattern is especially suited to fabrics with a high warp count. Textiles with a relatively low warp count (fewer than 60 or so warps per inch) are often decorated with two-face patterns. In two-face patterns the extra weft passes over or under both the odd-numbered and the even-numbered warps.

Brocade

We use the term brocade for patterns with extra-weft floats running parallel to the basic wefts. In single-face brocade, the floats appear only on the right side of the fabric. Good examples for single-face brocading are huipils from Almolonga, Cotzal, and San Antonio Aguas Calientes. Patterns woven in two-face brocading are typical for textiles from Concepción Chiquirichapa, San Martín Sacatepéquez, Nahualá, Chiché, and Chichicastenango. In two-face brocading, the extra weft goes to the underside of the material between surface floats. The float patterns on the underside produce a negative image of the pattern on the right side. Double-face brocading produces the same pattern on both sides of the fabric. This kind of pattern can be found on textiles from San Antonio Aguas Calientes.

Wrapped Weaves

In wrapped weaves the extra wefts are wrapped or looped around selected warps or groups of warps. Because of the wrapping, the extra wefts do not run parallel to the basic wefts. Each float is a little slanted. Single-face wrapped weave is typical for textiles from Colotenango and San Ildefonso Ixtahuacán. Weavers in Nebaj and Chajul also use several wrapping techniques. Two-face wrapped weave is the trademark of textiles from Zacualpa.

Twining

Twining requires two extra wefts to cross each other. The resulting weft floats slant a little and look much like wrapped weave. Some weavers in Colotenango and San Ildefonso Ixtahuacán do single-face twining to create horizontal bands.

For a detailed description of extra weft patterns and techniques involved in making these patterns, see chapter 20.

TAPESTRY EFFECTS

By tapestry effects we mean sections—mainly extra-weft patterns—of a textile where only wefts are visible. To create a weft surface on top of a ground fabric having a high warp count, several methods are employed:

Brocading

The standard brocading procedure, where each basic weft is followed by an extra weft (or a row of extra wefts) is not suited for producing a tapestry effect. The warps that hold down the extra wefts are clearly visible. In fact, they are chosen so as to contribute to the over-all pattern. To hide the warps, it is necessary to make at least two picks of the extra weft—each in its own shed—between picks of the basic weft. The warps that hold down the first pick of the extra weft are covered by the second pick, and the warps that hold down the second extra weft pick are hidden under the floats of the first pick. This method is very effective since extra wefts are much thicker than basic wefts (see chapter 20, "Huipil from San Antonio Aguas Calientes").

Women's belts in Nebaj are decorated with single-face patterns that have a weft surface (see chapter 20, "Woman's Belt from Nebaj"). Every basic weft is followed by four, six, or eight picks of extra wefts. Several methods are employed to superimpose the extra wefts over the ground wefts.

Wrapped Weave

Extra wefts that are wrapped around a warp (or a group of warps) always pass over that warp, thereby hiding it. Many huipils from Zacualpa have sections of two-face wrapping where no warps are visible. Huipils in Colotenango, San

Ildefonso Ixtahuacán, and Nebaj feature patterns in single-face wrapping that have a tapestry effect (see chapter 20).

Twining

Two-face or single-face twining also creates a weft surface. Weavers in Zacualpa, Colotenango, and San Ildefonso Ixtahuacán decorate huipils with twined bands.

Compound Weave

Weavers in Todos Santos and San Juan Atitán produce tapestrylike weft patterns by using a special weaving technique that is described in chapter 19 under "Looms with Additional Heddle and Shed Sticks" and in chapter 20 under "Huipil from Todos Santos Cuchumatán."

Weft Pile Weave

Weavers in Chichicastenango and Chiché decorate huipils and *servilletas* with the extensive use of extra wefts that form little loops. The end effect is much like tapestry. A description of the technique involved in producing this weave is described in chapter 19 under "Pattern Sticks for Two-Face Brocading" and in chapter 20 under "Huipil from Chiché."

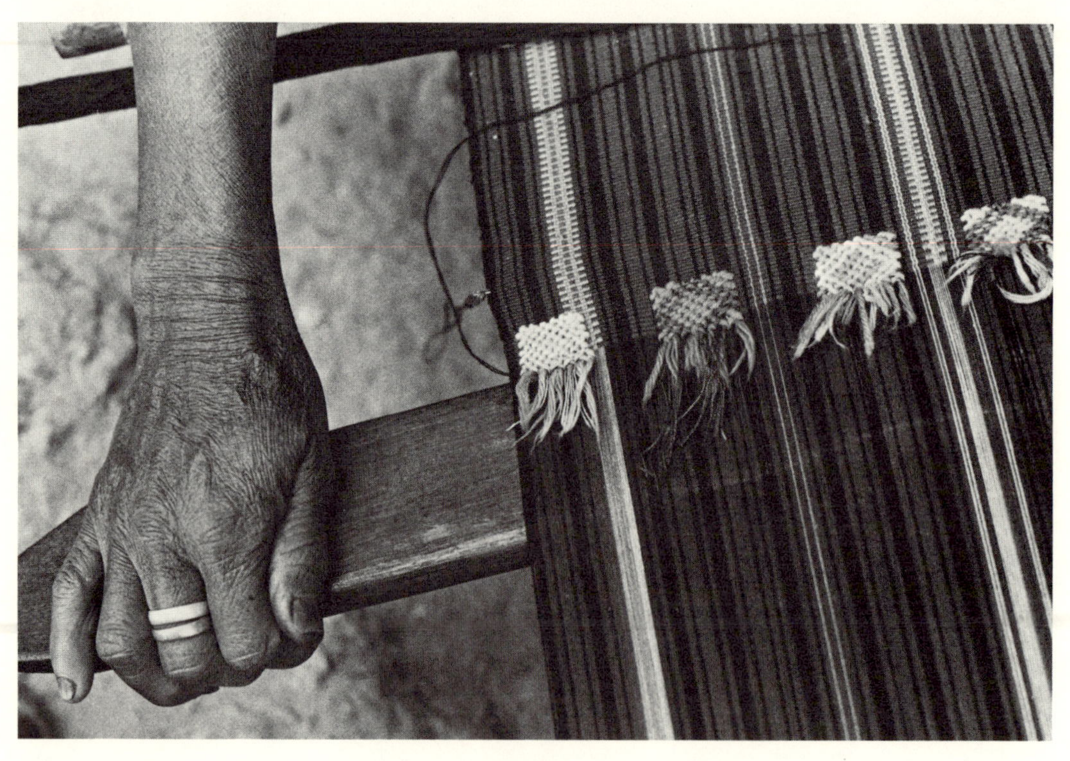

Plates 40 and 41. Weaving with extra wefts (single-face wrapping) in Colotenango. The weaver is working on a skirt.

18

»»»»»»»»»»«««««««««««

MATERIALS AND TOOLS FOR WEAVING WITH EXTRA WEFTS

MATERIAL

All commercial yarns listed in chapter 1 can be used as extra wefts. Two-ply yarn (Mish) and acrylic wool are very popular nowadays because they come in many colors and are less expensive than mercerized cotton or silk. Whichever type of yarn is chosen, the weaver always combines several plies to make an extra weft that is thicker than the wefts in the ground material.

TOOLS

To establish a shed for the extra weft, the weaver has to lift up selected warps or groups of warps. This can be done with her fingers, with a brocading sword or bone needle, or with the aid of pattern sticks and pattern heddles.

Wrapped weaving is done without the use of extra tools (see chapter 20, the huipils from Zacualpa, Colotenango, Nebaj, and Chajul). The same goes for weaving small discontinuous brocade patterns (see chapter 20, the huipils from Almolonga and Concepción Chiquirichapa). Women in San Antonio Aguas Calientes, however, often use bone needles to lift up the warps.

When brocading with continuous wefts, or if the pattern units are fairly large, weavers often employ *brocading swords* to establish the sheds for the extra wefts. Brocading swords are slender battens that are also used for weaving the join. To make a shed for the extra wefts, the weaver places the brocading sword over and under selected groups of warps and sets it on edge. Plates 42 and 43 show weavers from San Juan Atitán and Nahualá using brocading swords to make sheds for continuous extra wefts. Weavers in San Martín Sacatepéquez and Concepción Chiquirichapa also use brocading swords when doing two-face brocading (see chapter 20, "Huipil from San Martín Sacatepéquez").

In about half the towns we visited, weavers use

Plate 42. Two-face brocading in San Juan Atitán. The weaver uses a brocading sword to make a shed for the extra weft.

pattern heddles and/or pattern sticks for weaving extra-weft patterns. The operation of these devices will be discussed in the following chapter.

Besides the tools mentioned so far, bobbins are employed for weaving with continuous wefts. Weavers use their fingers to place discontinuous wefts in sheds.

Detailed descriptions of how the different patterns and designs are woven can be found in chapter 20.

Plate 43. Mother and daughter weaving in Nahualá. Both are doing two-face brocading with the aid of brocading swords. Both looms have two lease sticks.

Plate 44.

Weavers in Chajul. The loom at the right has two pattern sticks and two lease rods.

Plate 45.

A Chajul weaver inserts extra wefts under warps. The loose ends of the extra wefts remain on top of the fabric. The loom is equipped with one lease stick and two pattern sticks.

19

MODIFIED LOOMS

The standard backstrap loom is equipped for plain weaving. For other weaves (such as twill) and for the weaving of certain kinds of extra-weft patterns, extra heddles and/or shed sticks are added to the loom.

Weaves other than plain weave are used mainly to set decorative accents. Twill weaving is occasionally done in San Martín Sacatepéquez and Concepción Chiquirichapa. In San Juan Atitán and Todos Santos, weavers have developed an interesting compound weave that combines two sets of wefts with one set of warps. Weavers in Santiago Chimaltenango use a special weave derived from plain weaving to make weft stripes. All of these weaves are made with the aid of extra shed sticks and heddles.

In a number of towns, weavers use plain weave for the ground fabric and employ extra shed sticks and/or heddles to weave extra-weft patterns. If a shed for a brocading pattern is used more than once, it is worth the trouble to add a device to the loom that enables the weaver to reproduce the shed without having to count the warps anew. Extra shed sticks, also called pattern sticks, are easier to manipulate than extra heddles. Weavers in Aguacatán, Chajul, Cotzal, Nebaj, Chiché, and Chichicastenango use pattern sticks, but no pattern heddles, for weaving extra-weft patterns. Women in San Martín Sacatepéquez, Concepción Chiquirichapa, and San Antonio Aguas Calientes make use of both devices.

LOOMS WITH ADDITIONAL SHED STICKS (PATTERN STICKS)

The shed roll separates the odd- from the even-numbered warps. Either group of warps can be further divided by additional shed sticks. These sticks are smaller in diameter than the standard shed roll. Weavers in San Juan Atitán and Todos Santos often use brocading swords as pattern sticks. Pattern sticks are inserted into the odd- or even-numbered warps in back of the heddle. When not in use, they are moved behind the shed roll where they do not interfere with loom operations for plain weaving. To establish a shed, the pattern stick is moved close to the heddle. With the aid of

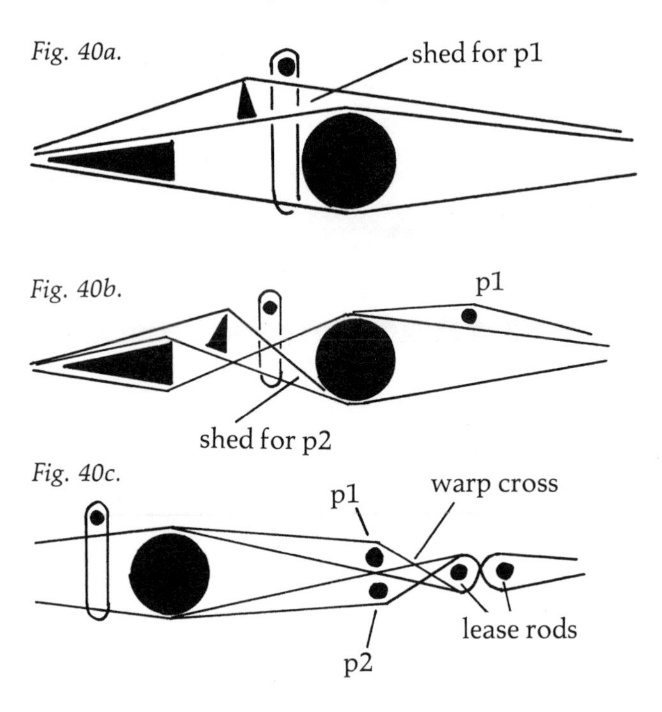

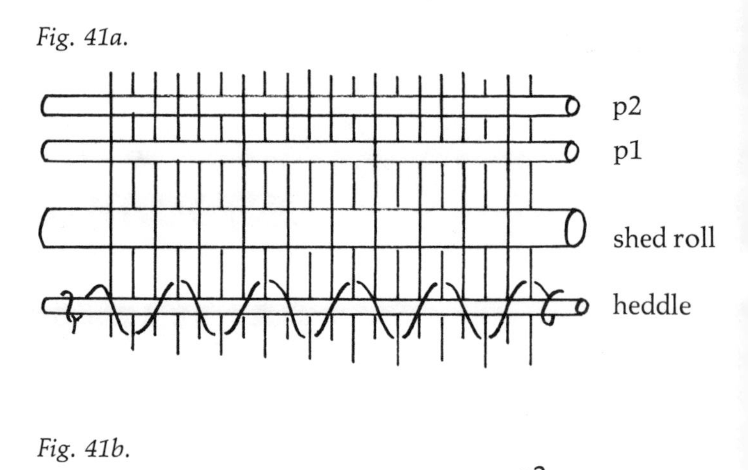

a brocading sword, the shed is then transferred to the front of the heddle. If the weaver wants to insert two pattern sticks into the odd- (or even-) numbered warps, she has to make sure there are no warp crosses between the sticks. Otherwise they cannot be moved past each other. We never saw more than four pattern sticks in one loom. Most weavers use pattern sticks in connection with single-face brocading. Weavers in Chichicastenango and Chiché use an ingenious method to do two-face brocading with the aid of two pattern sticks.

Pattern Sticks for Single-Face Brocading

Pattern sticks are especially suited for weaving horizontal bands. Such bands are produced by placing continuous extra wefts into sheds that are established by the pattern sticks. Good examples of this kind of weaving are the bands found on huipils from San Antonio Aguas Calientes (chapter 20, "Huipil from San Antonio Aguas Calientes") and from Cotzal (chapter 20, "*Servilleta* from Cotzal").

Weavers in San Antonio Aguas Calientes, Aguacatán, and Cotzal also employ pattern sticks when weaving with discontinuous extra wefts. The discontinuous wefts are either placed into sections of a shed produced by a pattern stick, or else the weaver selects some of the warps that are raised by the pattern stick to hold down the extra wefts (see figures 67 to 72). In the following, we describe weaving with pattern sticks as done in Cotzal. Weavers from this town decorate huipils and *servilletas* with rows of free-standing figures, separated by horizontal bands (plate 57). They use two pattern sticks for weaving the figures (birds, plant designs, geometric figures) and two more pattern sticks for the bands that run across the whole width of the fabric.

Putting Pattern Stick *p1* in Place

Antonia was weaving a panel for a huipil and had reached the point where she wanted to start the first row of bird figures. She put a basic weft in shed 1, battened it down, and left the batten in shed 1. To make a shed for the first pattern stick, Antonia inserted her brocading sword under every third warp that passed over the shed roll (warps 1, 7, 13 and so on). She picked up these warps close to the edge of the weaving, where the warps are in the right order. She moved the brocading sword close to the heddle and set it on edge to open the shed for pattern stick *p*1. The pattern stick was put into the shed *in back of the heddle,* as shown in figure 40*a*. With *p*1 in place, Antonia started the first row of extra wefts for the bird pattern. She moved the brocading sword close to the edge of the weaving and set it on edge, thereby raising every third one of the odd-numbered warps. Under some of these warps she inserted the extra wefts that were going to build up the bird pattern (see figure 68). When all the extra wefts were in place, Antonia removed the brocading sword from the loom and she pushed the pattern stick *p*1 behind the shed roll.

Putting Pattern Stick *p*2 in Place

Antonia made shed 2 and put a regular weft into it. The weft was battened down and the batten remained in the shed, with the even-numbered warps passing over the batten. With her brocading sword, Antonio picked up warps number 4, 10, 16, and so on. This was done close to the cross between the odd- and the even-numbered warps. Antonia then moved the brocading sword close to the heddle and set it on edge. In back of the heddle, she inserted pattern stick *p*2 under the even-numbered warps that were raised by the brocading sword (see figure 40*b*). Next, Antonia moved both *p*1 and *p*2 past the shed roll close to the lease rods (figure 40*c*). There, a cross is formed between the warps that go over *p*1 and the warps that pass over *p*2. Antonia checked the warp cross to make sure that the warps were in the correct sequence.

Figure 41*a* shows pattern sticks, shed roll, and heddle as seen from above, with the two pattern sticks positioned behind the shed roll.

Figure 41*b* shows which warps are raised by *p*1 and *p*2, assuming that *p*1 is employed while the batten is in shed 1, and *p*2 is used with the batten positioned in shed 2. The black squares indicate warps that are raised by the pattern sticks.

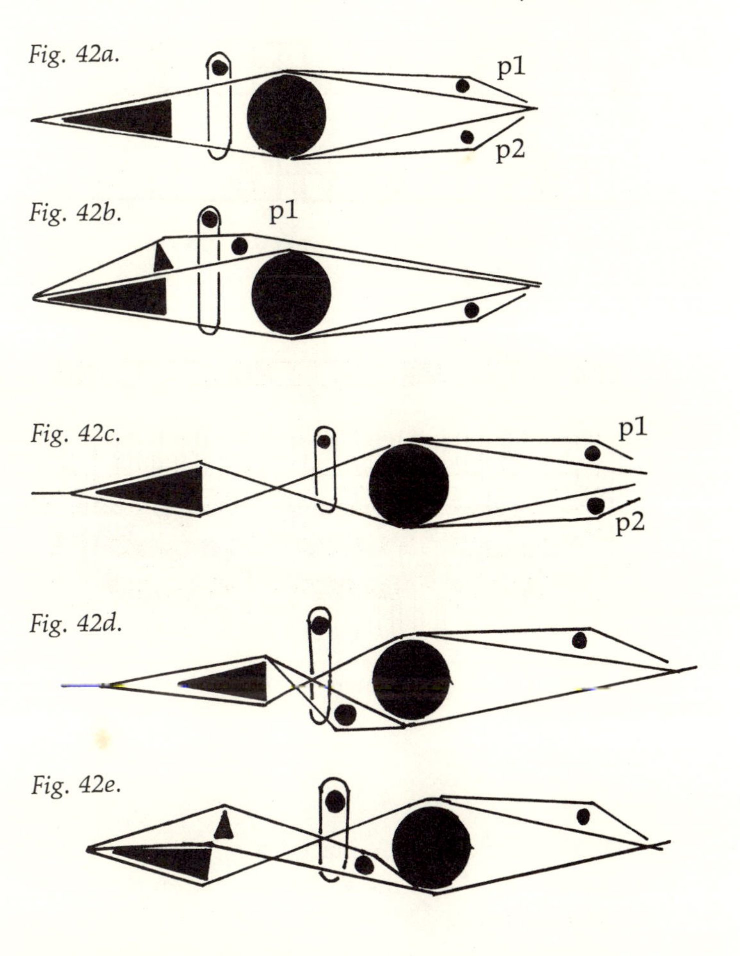

Weaving Procedure

With *p*2 in place Antonia moved the brocading sword close to the edge of the weaving, set it on edge, and put the extra wefts into their new sheds. From that point on she used the following weaving sequence:

1. She put the regular weft in shed 1 and battened it down. The batten remained in the shed (figure 42*a*).

2. Pattern stick *p*1 was moved close to the heddle. In front of the heddle, Antonia inserted the brocading sword into the shed that was made by *p*1. She set the brocading sword on edge (figure 42*b*) and placed the extra wefts under selected warps that passed over the brocading sword. After battening down the extra wefts, Antonia removed the brocading sword and pushed *p*1 behind the shed roll.

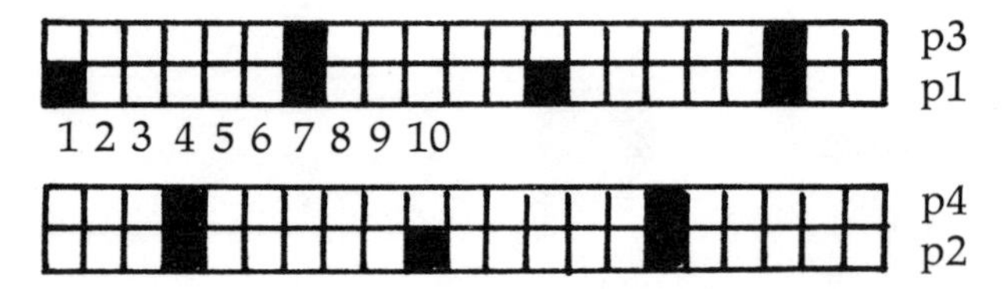

Fig. 43. Warps raised by *p*1, *p*2, *p*3, and *p*4.

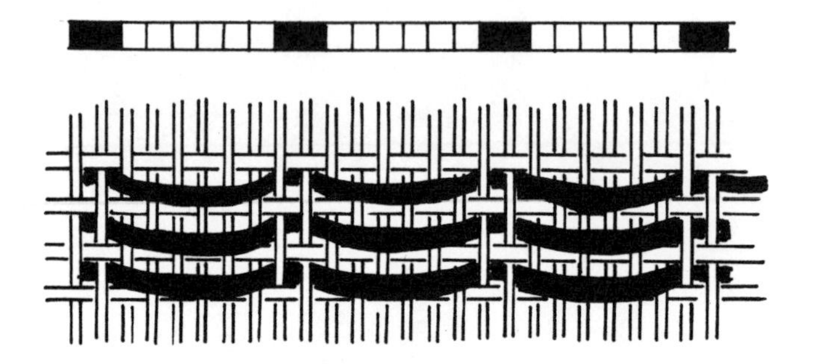

Fig. 44. Two-face brocading, Chichicastenango.

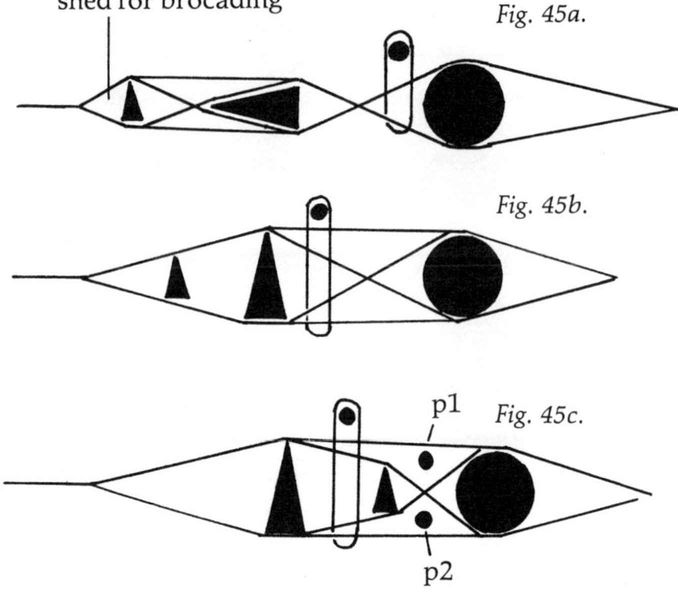

Establishing a shed for the pattern wefts.

3. She made shed 2, placed a regular weft in it and battened down the weft. The batten remained in shed 2 (figure 42*c*).

4. Pattern stick *p*2 was moved close to the heddle, the batten was also moved close to the heddle from the other side (figure 42*d*). In front of the heddle, Antonia inserted the brocading sword into the shed that was made by *p*2. She moved the brocading sword close to the edge of the weaving and set it on edge (figure 42*e*). The extra wefts were now placed under some of the warps that passed over the brocading sword. Antonia battened down the extra wefts with the brocading sword. Then she removed the brocading sword and pushed *p*2 behind the shed roll.

From the above description it is clear that the pattern sticks alone are not enough to establish the sheds for the extra wefts. Since the pattern sticks have to remain in back of the heddle, a brocading sword is needed to transfer the sheds to the front of the heddle.

When the bird pattern was finished, Antonia added two more pattern sticks to the loom. These additional sticks *p*3 and *p*4 were used for weaving horizontal bands. Figure 43 shows which warps are raised by *p*1, *p*2, *p*3, and *p*4. Between *p*1 and *p*3 no warp crosses occur, and the same is true for *p*2 and *p*4.

Pattern Sticks for Two-Face Brocading

In the area of our study, only weavers from Chiché and Chichicastenango use extra shed sticks for two-face brocading. In recent years they have given up their traditional patterns and replaced them with designs copied from embroidery-pattern books. These patterns are built up by short weft floats of uniform lengths. Only one shed is required for the extra wefts. The shed consists of two warps up, six warps down, repeated over the whole width of the fabric except for small margins along the side selvages. Extra wefts of different colors are inserted into different sections of this shed to establish the design (see chapter 20, "Huipil from Chiché").

Putting the Pattern Sticks in Place

The shed for the extra wefts is established with the aid of two pattern sticks. Pattern stick *p*1 raises the odd-numbered warps and *p*2 raises the even-numbered warps that pass over the extra wefts. To insert the pattern sticks into the warps, the following procedure is used:

The weaver places a basic weft in shed 2 and battens it down. The batten remains in shed 2, close to the heddle. With a brocading sword the weaver now establishes the shed that will be used for the pattern wefts. Close to the edge of the weaving she positions the brocading sword under two warps and over six warps in alternation (figure 45*a*).

This shed has to be transferred behind the heddle where the pattern sticks will be put in place. The weaver removes the batten from the loom and pushes the brocading sword close to the heddle. She sets the brocading sword on edge and inserts the batten alongside the brocading sword (figure 45*b*).

The brocading sword is taken out of the loom and reinserted into the shed in back of the heddle. The two pattern sticks *p*1 and *p*2 are inserted into the warp as shown in figure 45*c*.

These two sticks can be pushed behind the shed roll. The batten is still in the shed for the extra wefts in front of the heddle, and the weaver puts the first extra weft into this shed. From now on, picks of the basic weft alternate with picks of the pattern wefts.

Weaving Sequence

The weaver repeats the following four steps:

Step 1: Basic Weft in Shed 1

The pattern sticks are positioned behind the shed roll. The weaver makes shed 1 and puts a basic weft into the shed. She battens down the weft and removes the batten from the loom (figure 46*a*).

Step 2: Pattern Weft

To establish the shed for the extra weft, it is first necessary to make shed 2. The weaver raises the heddle, inserts the batten into shed 2 and battens down the warp cross. The batten is left in shed 2, just below the heddle. The two pattern sticks are brought forward from behind the shed roll and positioned close to the heddle, as shown in figure 46*b*.

Between the pattern sticks and the batten, the shed for the pattern wefts is formed. The weaver inserts the brocading sword into this shed. The batten is taken out of the loom. The brocading sword is set on edge and moved close to the heddle (figure 46*c*).

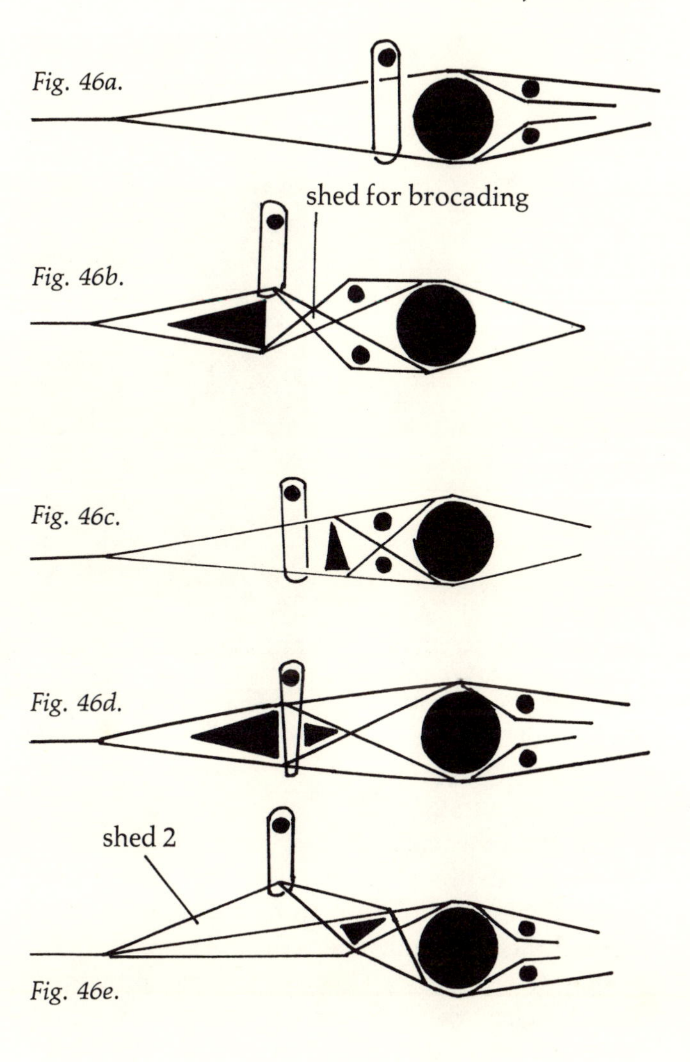

Fig. 46a. *Fig. 46b.* *Fig. 46c.* *Fig. 46d.* *Fig. 46e.*

Plate 46.

Pile-loop weaving in Chiché. With a hooking movement of her index finger, the weaver pulls the extra-weft floats from between the warps.

On the other side of the heddle, the batten is inserted into the shed that is held open by the brocading sword. The pattern sticks are pushed behind the shed roll. The weaver brings the brocading sword into a flat position, close to the heddle. To clear the shed for the pattern wefts, the weaver battens down the warp cross. The batten remains in the shed, close to the heddle and in a flat position (figure 46*d*).

The weaver puts a pick of the extra weft into the shed. Putting in the extra wefts requires a special technique that will be explained in chapter 20, "Huipil from Chiché."

Step 3: Basic Weft in Shed 2

The weaver pushes the brocading sword close to the shed roll and takes the batten out of the loom. She makes shed 2 by raising the heddle, and the basic weft is put into shed 2 as usual (figure 46*e*).

Step 4: Pattern Weft

The weaver moves the brocading sword close to the heddle and sets it on edge. The batten is taken out of shed 2 and reinserted into the pattern shed in front of the heddle. An extra weft is put into the shed and is battened down. The weaver takes the brocading sword and the batten out of the loom.

The above description of the weaving sequence contains a simplification. What is described as an extra weft is often a row of short wefts of different colors (see chapter 20, "Huipil from Chiché").

LOOMS WITH AN ADDITIONAL HEDDLE STICK (DOUBLE HEDDLE)

The standard heddle can easily be converted into a double heddle by adding an extra heddle stick (figure 47).

Lifting both heddle sticks together raises all the even-numbered warps. One half of the even-numbered warps is raised by lifting one of the heddle sticks.

Weavers in San Martín Sacatepéquez and Concepción Chiquirichapa use this setup to establish sheds for pattern wefts. For certain patterns they use a sequence of two basic wefts and one pattern

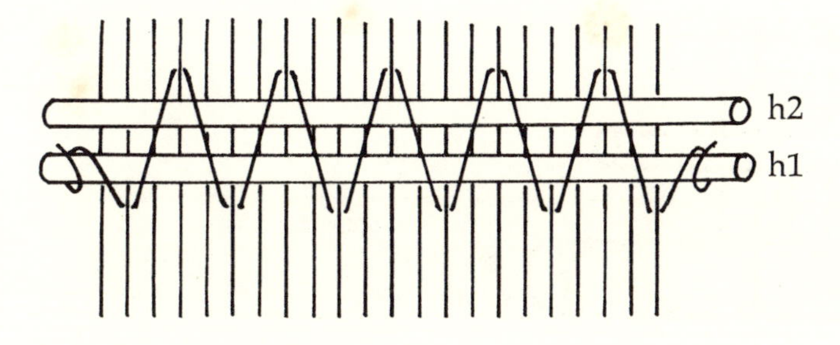

Fig. 47. Double heddle.

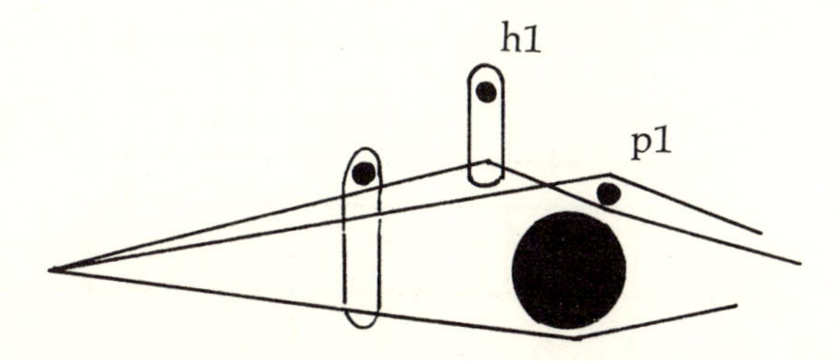

Fig. 48. Loom with pattern heddle and pattern stick, Concepción Chiquirichapa.

Fig. 49. Warps raised by *h*1 and *p*1.

weft. Every second basic weft—the one put in shed 2—is followed by an extra weft. With the batten still in shed 2, the weaver raises the front heddle stick *h*1 to make a shed for the extra weft. A brocading sword is put into the shed and set on edge. The weaver places the extra weft into this shed wherever the design requires it, and in between she brings the extra weft to the underside of the material.

All surface floats of the pattern weft are of equal lengths. Plate 84 (huipil from Cajolá, made in Concepción Chiquirichapa) shows a pattern made with the aid of an extra heddle stick.

LOOMS WITH ADDITIONAL HEDDLE AND SHED STICKS

Looms Equipped for Single-Face Brocading

Weavers in San Martín Sacatepéquez and Concepción Chiquirichapa decorate their textiles with two-face and single-face patterns. Both kinds can be seen in plate 65 (*servilleta* from Concepción Chiquirichapa). The birds are woven in a two-face technique with the aid of a brocading sword. Below and above the birds appear horizontal rows of extra weft floats. These floats are woven with the aid of a pattern stick and an extra heddle. Figure 48 shows the loom with the extra heddle *h*1 and the pattern stick *p*1. Figure 49 indicates which warps are raised by *h*1 and *p*1.

The sequence of weaving is as follows:

Step 1

The weaver puts a basic weft in shed 2 and battens it down.

Step 2

The weaver makes shed 1 and leaves the batten in the shed. She makes shed A for the extra weft and puts the extra weft loosely into it.

Step 3

The batten, still in shed 1, is set on edge and a basic weft is put in shed 1 and battened down.

This sequence is repeated three times, with the extra weft being placed in shed A. Then the weaver repeats the sequence three times, placing the extra weft in shed B. Three picks of the extra weft are put in shed A, three in shed B, and so on.

Shed A is made with the aid of pattern stick *p*1, shed B is made with the heddle *h*1. Stick *p*1 passes over three, under three of the odd-numbered warps. Heddle *h*1 controls the odd-numbered warps that pass under *p*1.

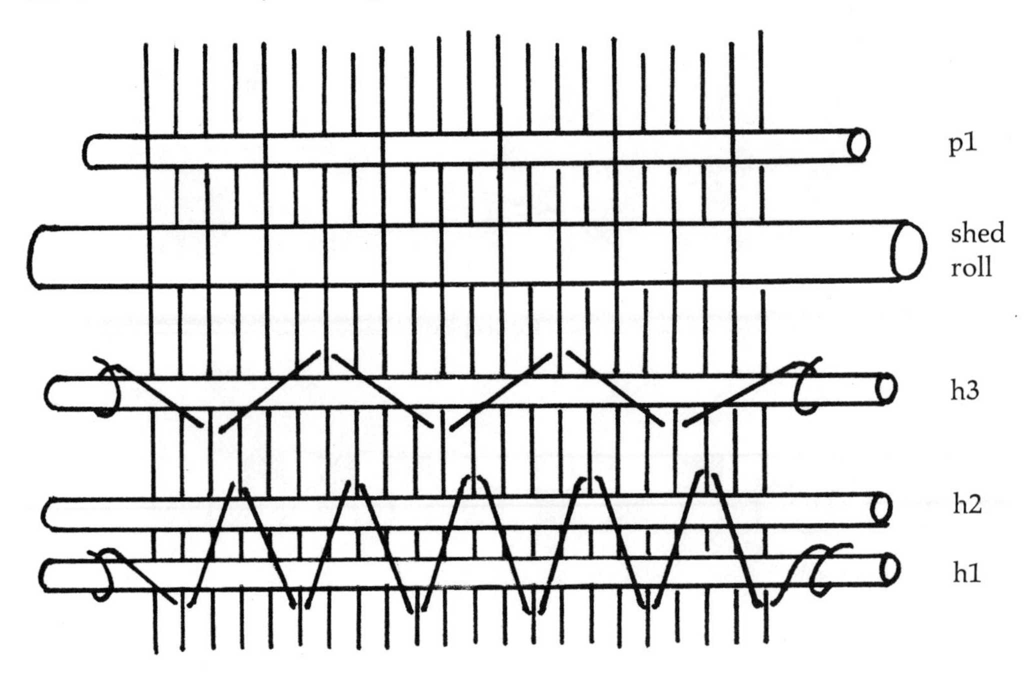

Fig. 50. Loom for twill weaving. Concepción Chiquirichapa.

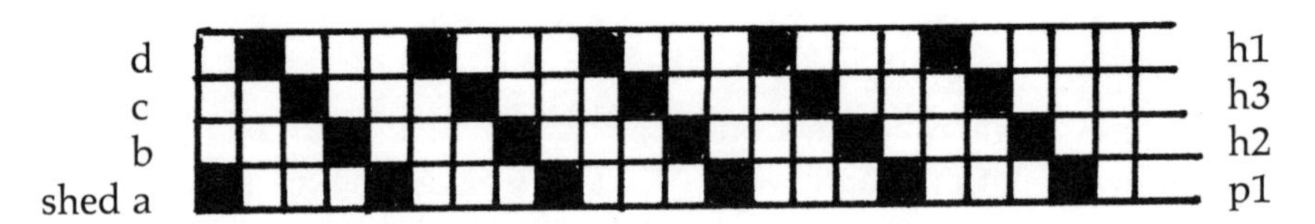

Fig. 51. Twill woven on loom shown in figure 50.

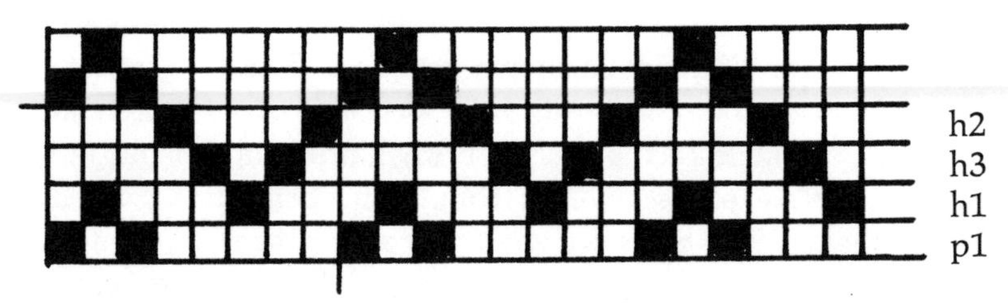

Fig. 52. Twill pattern from Concepción Chiquirichapa.

Looms for Twill Weaving

Weavers in Concepción Chiquirichapa weave twills that require four sheds. A twin heddle with two heddle sticks *h*1 and *h*2 divides the even-numbered warps. The odd-numbered warps are divided by an extra shed stick *p*1 and an extra heddle *h*3. Figure 50 shows a loom equipped for twill weaving. The extra shed stick *p*1 passes under every other odd-numbered warp. Heddle *h*3 controls the other half of the odd-numbered warps. Figure 51 shows the kind of twill that is woven with this loom.

Figure 52 shows another kind of twill. Here, stick *p*1 passes under two and over two odd-numbered warps, and heddle *h*3 controls the odd-numbered warps that pass under *p*1. We found both kinds of twill on huipils woven in the Cajolá style (see chapter 20, "Huipil in Cajolá Style").

The basic steps of weaving are the same for both kinds of twill. The weaver has to establish four different sheds, which we shall call sheds *a*, *b*, *c*, and *d*. For the twill pattern shown in figure 51, the sequence of weaving is as follows:

Shed a

The weaver moves shed stick *p*1 in front of the shed roll. With a brocading sword (indicated by a triangle in figure 53*a*) she further opens the shed that is made by *p*1. The brocading sword is moved close to heddle *h*3.

In front of the heddles, the batten is inserted into shed *a*. The weaver battens down the warp cross, sets the batten on edge, and puts a weft into the shed. The batten is then removed and stick *p*1 is pushed behind the shed roll.

Shed b

The weaver lifts heddle stick *h*2 and pushes down heddle stick *h*1 at the same time. She inserts the brocading sword under the warps that are controlled by *h*2 and sets the sword on edge (figure 53*b*).

In front of the heddles, she places the batten into shed *b*. The warp cross is battened down, the batten is set on edge and a weft pick is put into shed *b*.

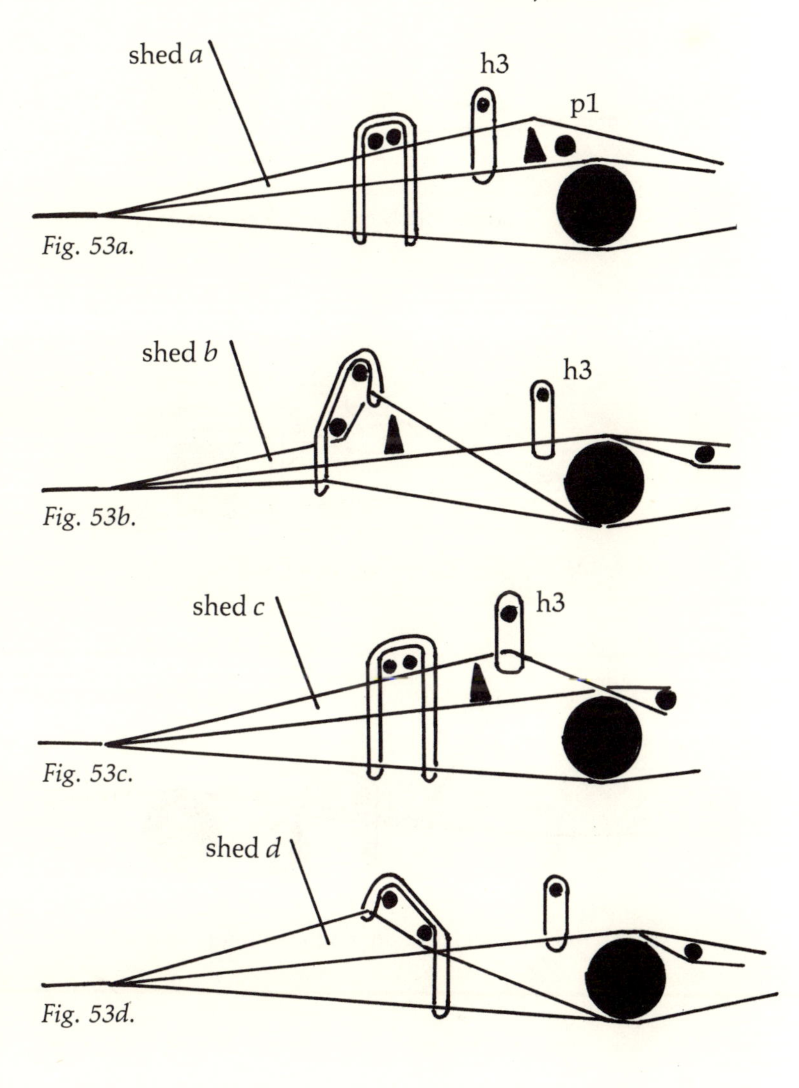

Fig. 53a.

Fig. 53b.

Fig. 53c.

Fig. 53d.

The weaver then takes batten and brocading sword out of the loom.

Shed c

The weaver lifts heddle stick *h*3 and places the brocading sword under the warps controlled by *h*3, as shown in figure 53*c*.

Again, the shed is transferred to the front of the heddles where the batten is put into the shed. The weaver battens down the warp cross, sets the batten on edge, and puts a weft into shed *c*. Batten and brocading sword are then taken out of the loom.

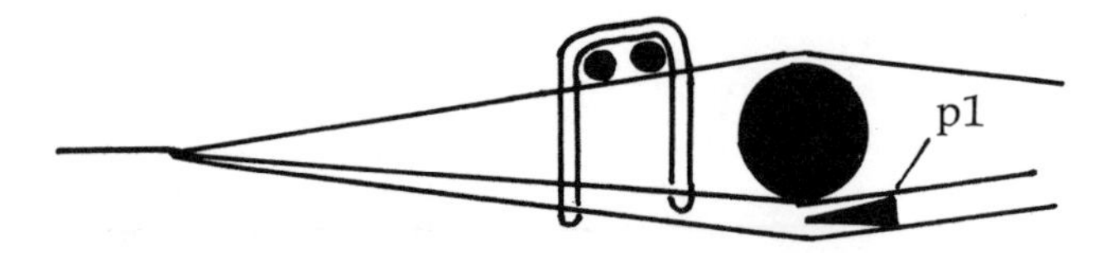

Fig. 54a. Loom from Santiago Chimaltenango.

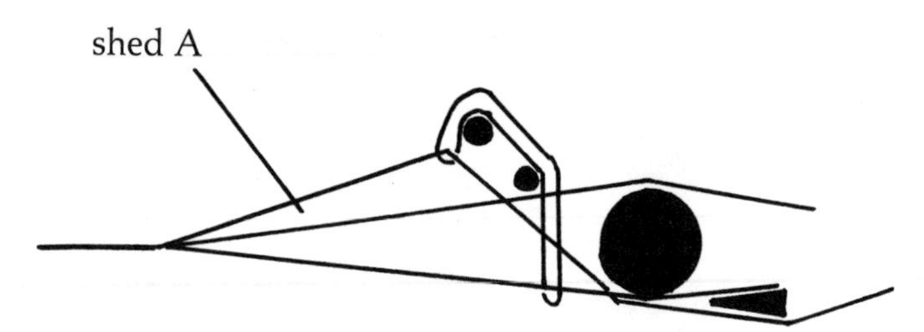

Fig. 54b. Loom from Santiago Chimaltenango.

Fig. 54c. Loom from Santiago Chimaltenango.

Fig. 55. Weft stripe pattern, Santiago Chimaltenango.

Shed d

The weaver lifts heddle stick *h*1 (figure 53*d*) and inserts the batten under the warps controlled by *h*1. She battens down the warp cross and puts a weft into shed *d*.

Looms for Weaving Weft Stripes (Santiago Chimaltenango)

Weavers in Santiago Chimaltenango decorate their huipils with both warp and weft stripes. Since the warps are closely packed, a special technique is employed to make the weft stripes stand out. The weaver converts the heddle into a double heddle. Each heddle stick controls one half of the even-numbered warps. A brocading sword, functioning as shed stick *p*1, separates the two groups of heddle-controlled warps as shown in figure 54*a*.

For the weaving of the weft stripes, two sheds are used in turns. We shall call them shed A and shed B.

Shed A

The weaver lifts heddle stick *h*1 (figure 54*b*) and inserts the batten under the warps controlled by *h*1. She battens down the warp cross, sets the batten on edge, and puts a stripe weft into shed A. The weft is battened down and the weaver removes the batten.

Shed B

Shed stick *p*1 is brought forward close to the heddle and set on edge (figure 54*c*), thereby opening shed B which is the countershed to shed A. The batten is inserted into shed B in front of the heddle. The weaver battens down the warp cross, sets the batten on edge and puts the next stripe weft into the shed. The weft is battened down. Since all the odd-numbered warps and half of the even-numbered warps pass over this weft, it is only visible on the underside of the material.

Altogether, ten picks of stripe wefts are made in a row. Then the weaver returns to plain weaving until she starts a new series of weft stripes. Shed

Plate 47. Section of a huipil from San Juan Atitán.

stick *p*1 and the extra heddle stick are kept in the loom during plain weaving.

Figure 55 shows the pattern produced by the stripe wefts. A huipil from Chimaltenango can be seen in plate 82.

Loom for a Compound Weave (San Juan Atitán and Todos Santos)

Weavers in Todos Santos and San Juan Atitán employ a compound weave to create patterns with a weft surface. Plate 47 shows a section of a huipil from San Juan Atitán. The horizontal bands that are raised above the white areas of plain weave are done in a compound weave. What on first glance appears to be single-face brocade is actually a weave with a more complex structure. For this kind of weave two sets of wefts are used. The surface wefts appear on the right side of the fabric, completely covering the warps. The second set of wefts—we shall call them ground wefts—appears on the underside of the fabric. One set of warps interlaces with the surface wefts, a second set interlaces with the ground wefts, and a third set of warps connects the surface wefts with the ground wefts. Since this weave has more than the customary two sets of complementary elements (warps and wefts), we refer to it as a compound weave.

Figure 56 shows the transition from plain to compound weave as it appears on the underside of the fabric. The basic weft in plain weaving becomes the ground weft in the compound weave.

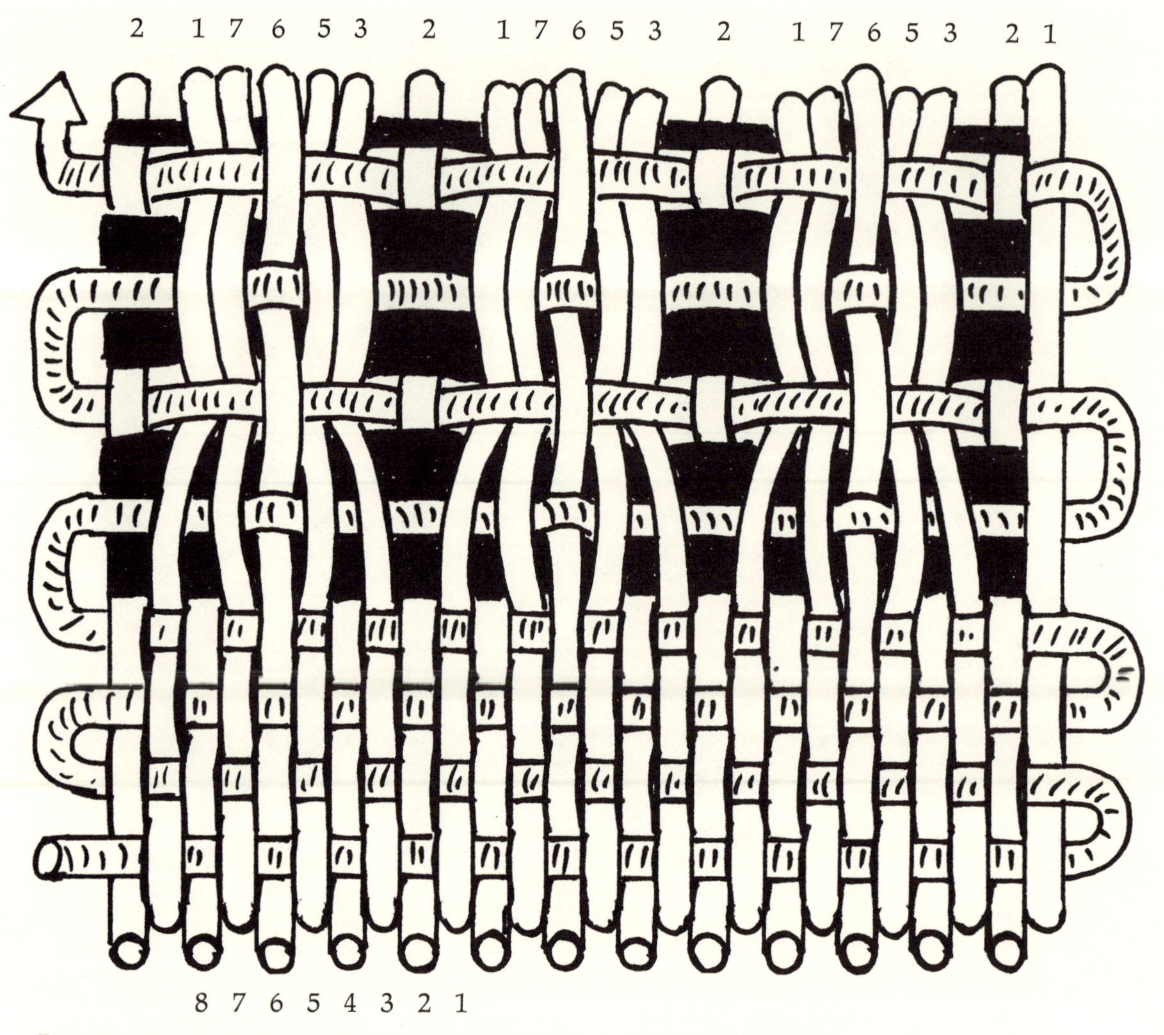

Fig. 56. Transition from plain to compound weave, wrong side of fabric.

Each black band in figure 56 represents several picks of surface wefts, battened closely together (how these wefts interlock with the warps is shown in figure 57). From figure 56 we can conclude that the basic unit of construction for the compound weave consists of eight warps, two ground wefts and a number of surface wefts. The warps are numbered from 1 to 8, and all warps with the same number interlace in the same way with the ground and/or surface wefts. Out of the eight warps that go into the basic unit of construction, five warps (numbers 1, 3, 5, 6, and 7) interlock only with the ground wefts. One warp (number 2) interlocks with both the ground and the surface wefts, connecting the two sets of wefts. Two warps (numbers 4 and 8) do not appear on the underside of the fabric. They interlock only with the surface wefts.

In figure 57, illustration *a* shows the transition from plain to compound weave as it appears on the right side of the fabric. In the diagram the surface wefts (black) are loosely spaced, but in the actual weaving they are close together. As shown in figure

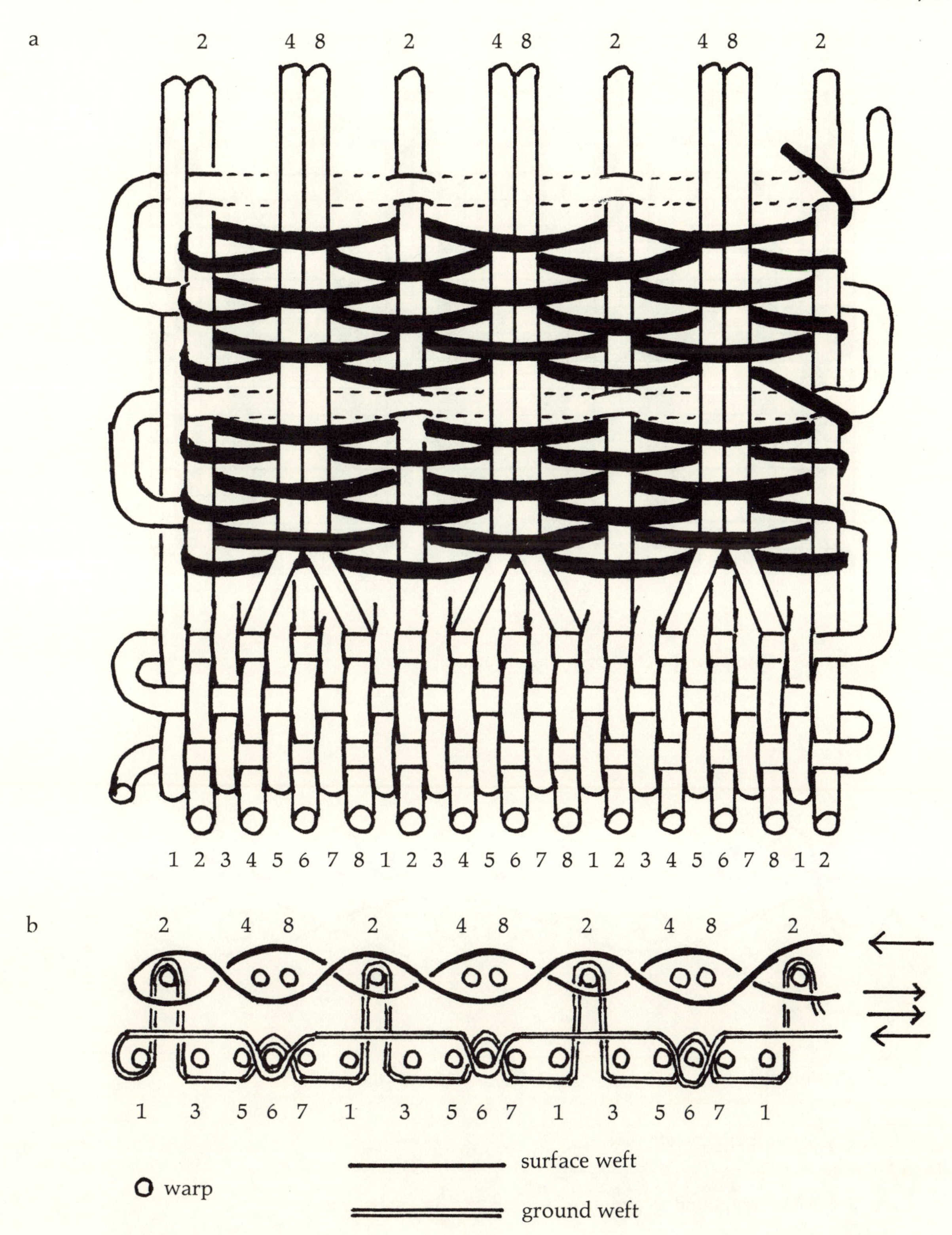

Fig. 57. Compound weave: *a*, transition from plain to compound weave, right side; *b*, cross section of compound weave, cut parallel to the wefts.

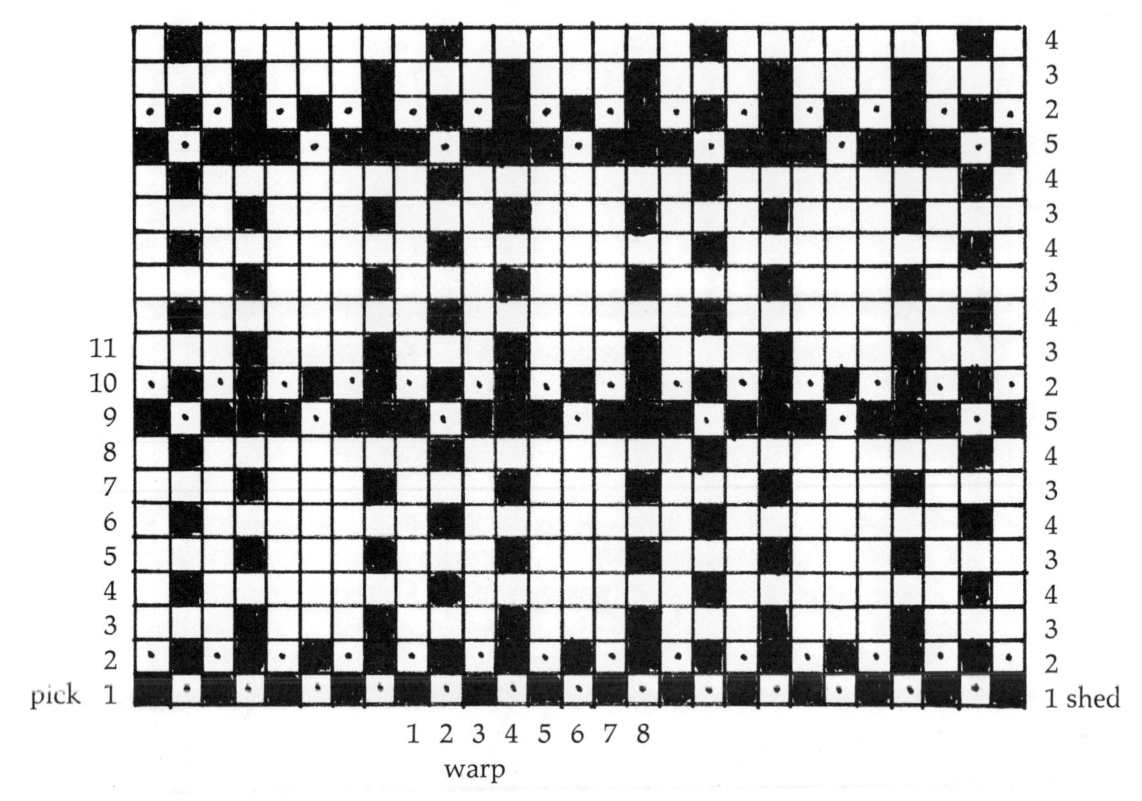

Fig. 58. Compound weave, weaving sequence.

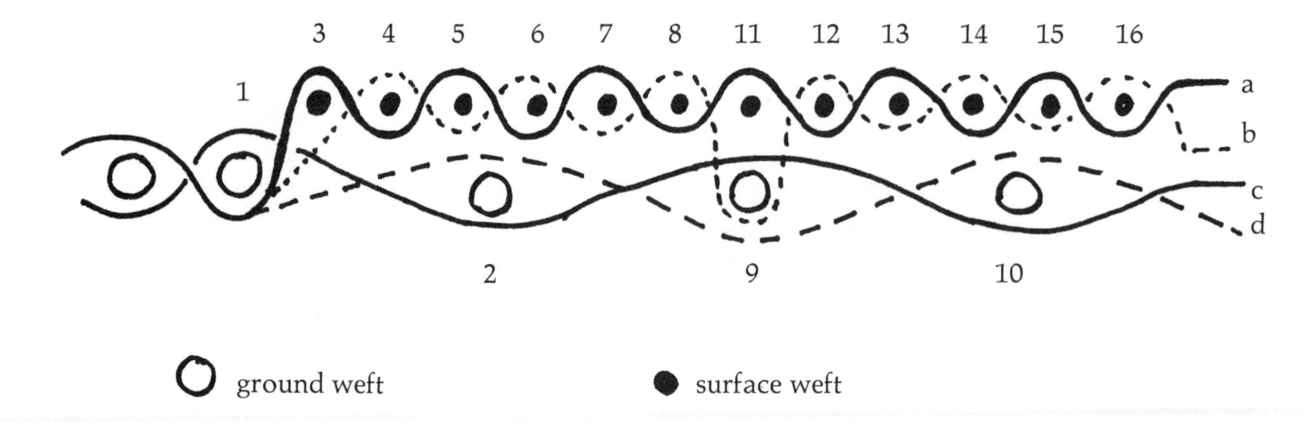

Fig. 59. Compound weave, cross section parallel to the warps.

57*a*, warps 4 and 8 are pushed close together by the surface wefts. These two warps do not interlace with the ground wefts, and removal of the surface wefts would leave them bare. Warp 2 interlaces with both ground and surface wefts. Warps 1, 3, 5, 6, and 7 are omitted in the diagram, since they do not interlace with the surface wefts. Because the surface wefts are much thinner than the ground wefts, the weaver makes six, eight, or more picks of the surface weft for two picks of ground weft. Every second ground weft passes over warps number 2, dividing the surface wefts into groups of six. In the actual weaving, these groups appear as cords (see plate 47).

In figure 57, illustration *b* represents a cut through the fabric parallel with the wefts. The cross sections of the warps appear as circles. Both ground and surface wefts are shown interlocking with the warps. Only two picks of the surface weft are shown. In the actual weaving, six or more picks of surface weft are made between picks of the ground weft. The weaving sequence, in form of a block diagram, is given in figure 58.

Each horizontal row of squares represents one weft pick. The black squares are warps that pass over the weft. *White squares* represent the *surface weft* passing over a warp. *White squares with a dot* represent the *ground weft.*

The first two picks are done in plain weave. Compound weave starts with pick number 3. After six picks of surface weft, there follow two picks of ground weft. This sequence—six surface wefts, two ground wefts—is repeated until the weaver switches to plain weaving.

At the right-hand side of the diagram appear numbers indicating what shed is used for each pick. Sheds 1 and 2 are the sheds used in plain weaving. Sheds 3 and 4 accommodate the surface wefts. Shed 5, which can be interpreted as shed 1 plus shed 3, is needed for the first ground weft after six surface wefts. Next comes shed 2 for the ground weft. The surface weft is placed in sheds 3 and 4 in alternation, the ground weft in sheds 5 and 2.

While indicating the weaving sequence, figure 58 does not show the position of surface and ground wefts in the woven material. There the ground wefts are positioned *below* the surface wefts. This is shown in figure 59, which represents a cross section of the fabric parallel with the warps.

Each number in figure 59 indicates a weft pick and corresponds to the numbers at the left in figure 58. The cross sections of wefts appear as circles. There are four lines representing warps. Line *a* represents the warps that are raised to establish shed 3. Line *b* represents warps that are raised to establish shed 4. Line *c* represents all the odd-numbered warps (shed 1) and line *d* represents the even-numbered warps that are not employed for making sheds 3 and 4. Shed 5 is produced by raising warps *a* and *c*. The diagram shows how, because of the position of the warp crosses, the ground wefts go to the underside of the fabric.

The weaving sequence as outlined above can be modified by starting the compound weave with shed 4 instead of shed 3 or by increasing or decreasing the number of surface weft picks between picks of the ground weft. In any event, the weaver has to establish the three extra sheds: sheds 3, 4, and 5.

Modifying the Loom

In order to make sheds 3, 4, and 5, the loom has to be equipped with a double heddle and two extra shed sticks. We observed a weaver in San Juan Atitán adding these parts to her loom. First she added an extra heddle stick to the standard heddle as follows: With the end of a spindle she pulled forward every other one of the heddle loops hanging down in front of the heddle. Under these loops she inserted the extra heddle stick *h*1 (figure 60).

Next she added shed sticks 1 and 2 (two brocading swords) to the loom as follows: She lifted up heddle *h*1 so high that *all* the even-numbered warps were raised above the odd-numbered warps (figure 61*a*). Shed stick *s*1 was placed under the warps that are controlled by the loops in back of *h*2.

Next the weaver moved *h*1 and *h*2 close together and set *s*1 on edge (figure 61*b*). Shed stick *s*2 was placed behind *s*1 (under the odd-numbered warps and the warps that pass over *s*1) and pushed behind the shed roll.

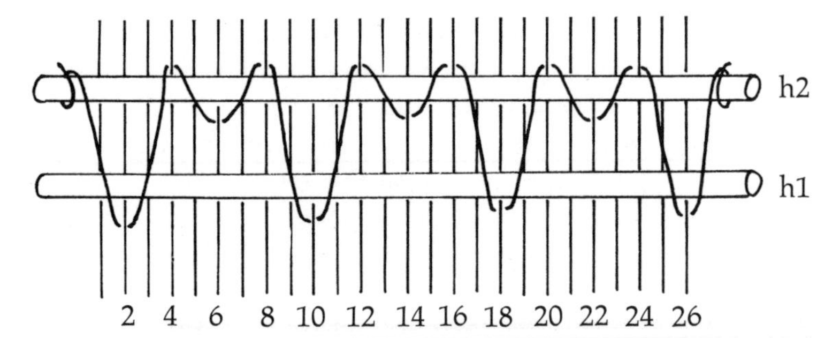

Fig. 60. Double heddle. San Juan Atitán.

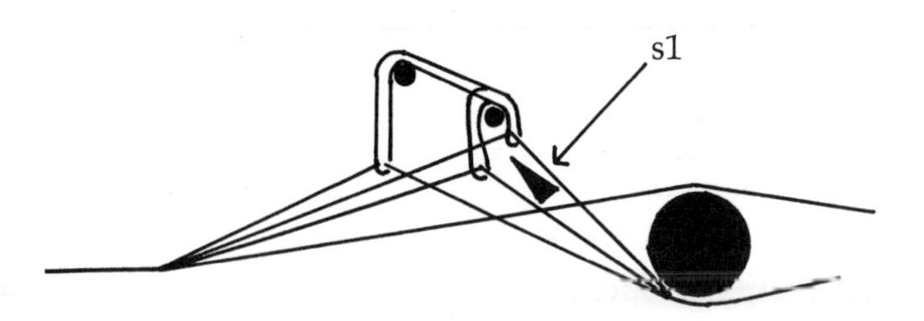

Fig. 61a. Adding shed stick *s*1 to the loom.

Fig. 61b. Adding shed stick *s*2 to the loom.

Weaving Procedure

The weaving sequence is as follows: Six or more picks of surface weft alternating in sheds 3 and 4, one pick of ground weft in shed 5, one pick of ground weft in shed 2.

Shed 3

The weaver sets shed stick *s*1 on edge and moves it close to the heddles. She inserts the batten in the shed that is made by *s*1 in front of the heddle. The warp cross is battened down, and the batten is set on edge (figure 62*a*).

The weaver puts a surface weft into the shed. The weft is not pulled straight across, it is set in position as shown in figure 62*b*. If put in straight, the surface weft would pull the warps together after being battened down, which would make the fabric narrower. This narrowing effect can never be completely avoided when doing this kind of weave. Shed stick *s*1 returns to the flat position and the batten is removed from the loom.

Shed 4

Lifting up heddle stick *h*1 and pressing down *h*2 establishes shed 4. The weaver inserts the batten into shed 4 and battens down the warp cross, thereby battening down the last surface weft that had been placed in shed 3. The batten is set on edge (figure 62*c*) and a surface weft is placed in the shed in the same way as shown in figure 62*b*. The batten is removed from the shed.

Shed 5

Shed stick *s*2 is brought forward from behind the shed roll and set on edge (figure 62*d*). The batten is inserted into the shed established by *s*2 in front of the heddle and the warp cross is battened down. The weaver puts a ground weft into the shed and battens it down.

Unlike surface wefts, ground wefts are put straight across into their sheds. The weaver pushes stick *s*2 behind the shed roll and takes the batten out of the loom.

Next comes a ground weft in shed 2, followed by a surface weft in shed 3. Battening down the warp cross of shed 3 also battens down the last surface weft that had been put in shed 4.

On huipils from San Juan Atitán, decorative red bands with a weft surface alternate with plain weaving. For the plain weaving, *s*1 has to be taken out of the loom. Shed stick *s*2 remains positioned behind the shed roll. Both heddle sticks remain in the loom and are lifted together to establish shed 2. When the compound weave starts again, *s*1 has to be put in place. The weaver places the batten in shed 2 and sets it on edge. She moves shed stick *s*2 in front of the shed roll and sets it on edge, too. Shed stick *s*1 can now be put back in place, as shown in figure 62*e*.

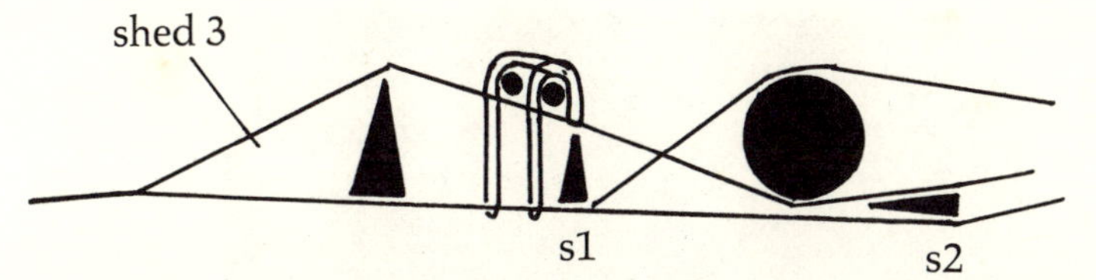

Fig. 62a. Compound weave, shed 3.

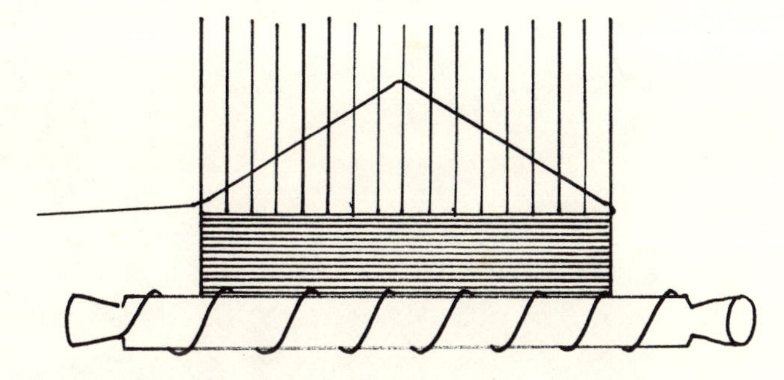

Fig. 62b. Compound weave, position of surface weft before it is battened down.

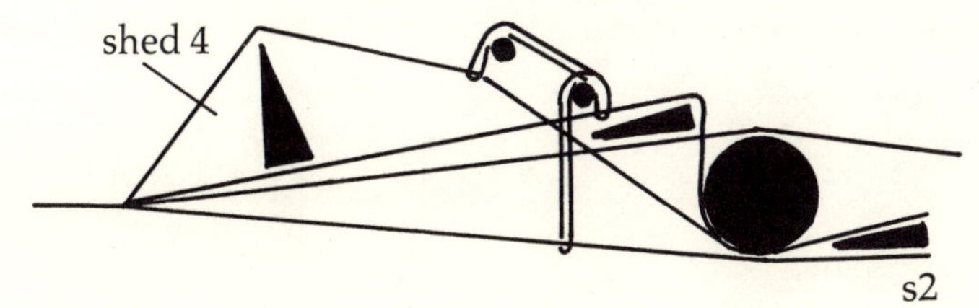

Fig. 62c. Compound weave, shed 4.

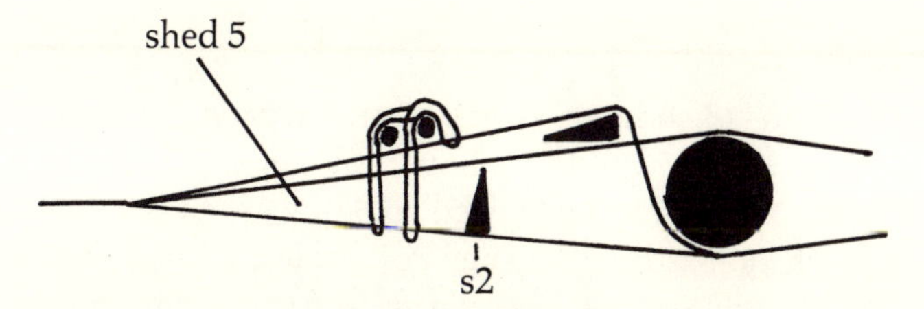

Fig. 62d. Compound weave, shed 5.

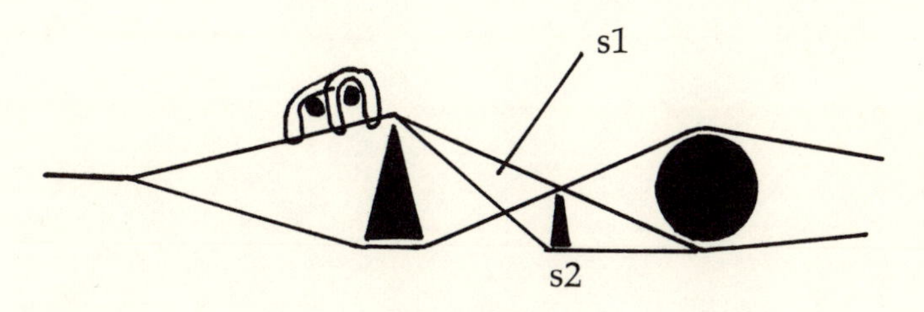

Fig. 62e. Putting *s*1 back in place.

Plate 48. Todos Santos girl weaving material for shirt collars.

Plate 48 shows a Todos Santos weaver at work. From left to right the following loom sticks can be seen: two lease sticks tied together, shed stick *s*2 (with pointed ends), shed roll (round), shed stick *s*1, heddle stick *h*2, heddle stick *h*1, batten, tenter (under the edge of the weaving), end bar, and cloth bar.

Weavers in San Juan Atitán employ the compound weave to create decorative bands. In Todos Santos women like to experiment with the possibilities of the compound weave. They add extra wefts between picks of the surface wefts, as we shall discuss in chapter 20, "Huipil from Todos Santos Cuchumatán." The latest fashion in Todos Santos weaving is to build up tapestry patterns with discontinuous surface wefts.

20

WEAVING TECHNIQUES AND PATTERNS

From our collection of textiles we selected twenty-one items to illustrate different weaving techniques and patterns. The first two textiles discussed, a *delantera* from San Sebastián Huehuetenango and a huipil from San Pedro Necta, have stripe patterns. Single-face brocading is used to decorate a huipil from Almolonga, a headband from Aguacatán, and a *servilleta* from Cotzal. A huipil from San Antonio Aguas Calientes shows both single-face and double-face brocading. The first Nebaj huipil has a weft surface and is decorated with weft stripes and single-face patterns. Two-face brocading is featured on textiles from San Martín Sacatepéquez, Concepción Chiquirichapa, and Nahualá. The huipil from Chiché has sections that are woven in a weft pile weave. Two-face wrapped weave and twining are the outstanding features of a Zacualpa huipil. Huipils from Colotenango, Nebaj, and Chajul show patterns done in single-face wrapped weave and single-face brocade. The same techniques are used to decorate a belt from Nebaj. Belts from Santiago Chimaltenango and Totonicapán have warp float patterns. A huipil woven in Cajolá style serves as an example of twill weaving, and a huipil from Todos Santos features an unusual compound weave.

DELANTERA FROM SAN SEBASTIAN HUEHUETENANGO

Special feature: Check pattern in black and white wool (plate 50).

Many men living near the town of San Sebastián Huehuetenango still wear a *"lanter"* (from the Spanish word *delantera*), a wool blanket that is wrapped around the hips (see plate 13). In other places (Nahualá and towns around Lake Atitlán) these blankets are woven on foot looms. In San Sebastián, however, there are very few foot-loom weavers, and woolen articles are also woven on the backstrap loom. Plate 49 shows weaving of the *delantera*. Plate 50 shows a close-up of the piece. The piece was bought unfinished, together with the loom.

Plate 49. Weaving a *delantera* in San Sebastián Huehuetenango.

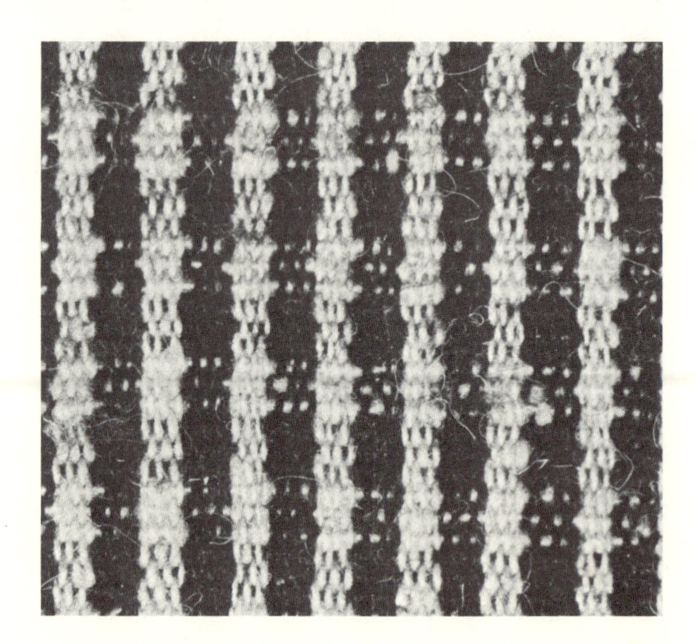

Plate 50. Section of a *delantera* from San Sebastián Huehuetenango.

Weaving Procedures

The weaver told us that one pound of white and one pound of black raw wool would be required for the *delantera.* Spinning of the wool was done with a spindle to get a fine thread. Warping was done on five sticks, using the *cadena* technique for the black and white stripes. Warping took less than two hours. Three circuits of white yarn are followed by 3 circuits of black yarn, and so on. Altogether, 288 circuits were made. The warps pass over the end bars on the loom, since the *delantera* is woven with end fringes. On the loom the material is 19 inches wide and 70 inches long. The thread count is 33 warps and 12 wefts per inch. The weft stripes, also in black and white, do not show up very well. Three picks of white yarn are followed by three picks of black yarn in alternation.

HUIPIL FROM SAN PEDRO NECTA

Special feature: Warp stripes (plates 51, 52).

Huipils in San Pedro Necta are made from one panel with four selvages. The warp stripes run horizontally on the finished huipil. This is unusual. There are also few other places in Guatemala where huipils are made from just one panel. The huipil is 20 inches long and 30½ inches wide. It is worn over the skirt. Warps and wefts consist of paired singles. For the warp, white (no. 10?), red (no. 8) and finer yarn in blue, green, and yellow are used. The latter three colors are used sparingly to set accents. The weft yarn is white. The warp count is 44 pairs per inch, the weft count 17 pairs per inch. The main feature of this huipil is the arrangement of the warp stripes. Figure 63 shows in detail one of the stripe patterns. Warping the stripes must have been quite tedious, because several hundred knots had to be made to tie warps of different colors together. A *cadena* technique (which makes tying of the warps unnecessary) was not used.

Nowadays, this kind of huipil is worn mostly by older women. It lacks the novelty of the extra-weft patterns that have been added to the traditional design in recent years. Most younger weavers in San Pedro have learned to make extra-weft patterns in the style of weavers from Colotenango and San Ildefonso Ixtahuacán.

HUIPIL FROM ALMOLONGA

Special features: Single-face brocading, warp and weft stripes (plates 53, 54).

Huipils in Almolonga are made from two panels. Each panel is woven separately, but has only three selvages. One end of the material is cut when it is taken from the loom. This way the weaver does not have to make a join.

The huipil is 41 inches long and 31½ inches wide. The lower part is worn under the skirt. The upper part is completely covered with extra-weft patterns, done in single-face brocade. Holes for the head and arms are seamed with velvet ribbons. The huipil has red and white warp stripes, visible only on the lower part of the garment. Four white warps alternate with four red warps. A white warp consists of two singles, a red warp of three singles. The red yarn is no. 20, the white yarn probably no. 16. The warps are spaced rather loosely, thread count being about 38 warps per inch. In spite of the low warp count, the fabric is firm and heavy, because very strong wefts are used. In the lower part of the huipil, about thirty picks of white weft alternate with two picks of red weft. The white weft consists of twelve singles of no. 20 yarn, the red weft of four singles, also no. 20. In the upper part of the huipil, the wefts are black (faded to a dark blue). One weft consists of four two-ply yarns. The weft count is 17 wefts per inch for the white wefts and 22 wefts per inch for the black wefts.

Extra-Weft Pattern

Weavers in Almolonga combine a few simple geometrical forms to create abstract patterns. Colors are carefully chosen and contribute, together with the pattern, to the "modern" appearance that is

Plates 51 and 52. Huipil from San Pedro Necta.

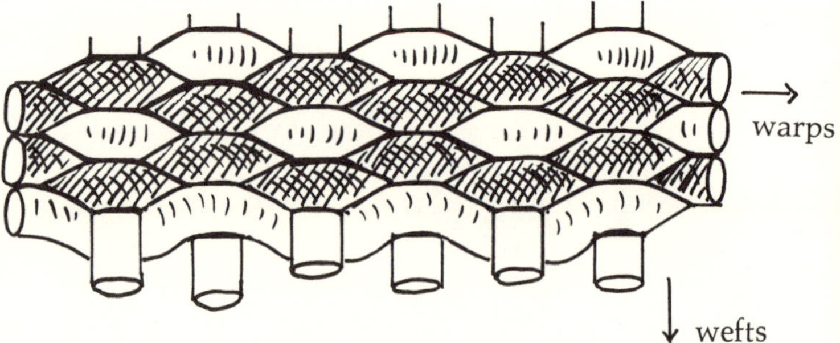

Fig. 63. Warp stripe, San Pedro Necta.

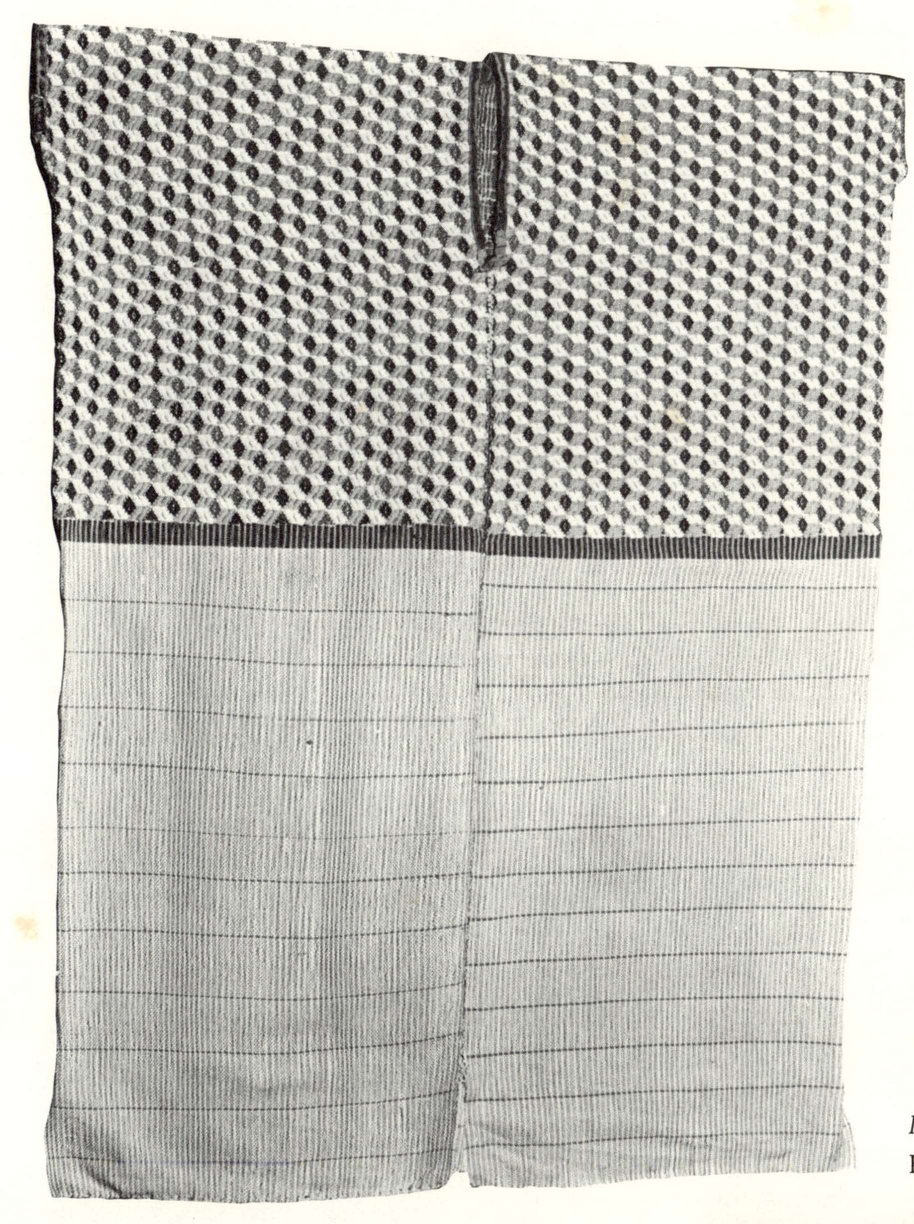

Plates 53 and 54.
Huipil from Almolonga.

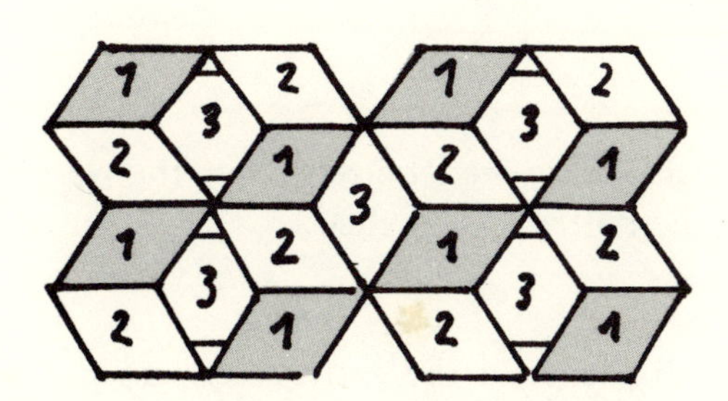

Fig. 64. Extra-weft pattern, basic elements.

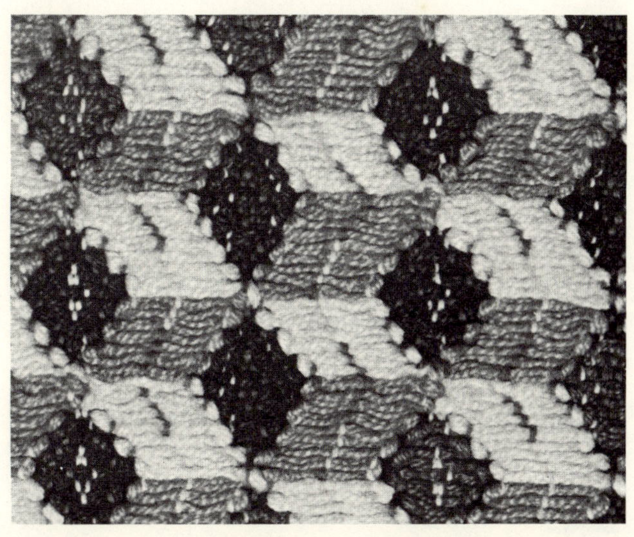

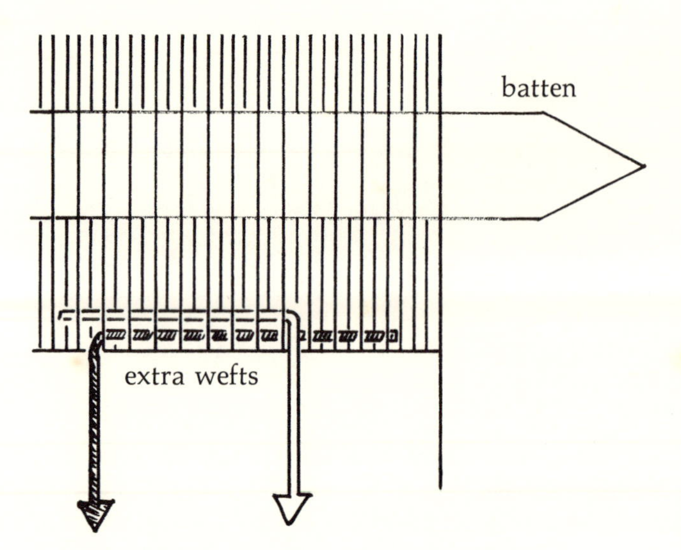

Fig. 65. Starting the extra wefts, Almolonga.

typical for huipils from this town. On the huipil pictured in plates 53 and 54, the pattern consists of three geometrical forms, each one repeated eleven times across the width of each panel.

The three basic elements of the pattern are shown in figure 64. Element 1 is done in orange yarn, element 2 is white and element 3 is green, red, or mauve. Three strands of Sedalina are combined for the extra wefts.

When weaving such patterns as the one described here, weavers always use a huipil, or some other textile, as a model. The number of warps and the arrangement of the warp stripes is related to the extra-weft patterns. In order to reproduce a pattern precisely, the weaver uses the warp stripes as markers. For the upper part of the huipil, dark-blue wefts are used instead of white ones. The white warps show up better against the dark-blue wefts. This is very convenient for weaving the extra-weft pattern when it is necessary to lift up specific warps.

The pattern starts with 11 white and 11 orange extra wefts. The weaver cuts 22 lengths of brocading yarn, each one about 6 inches long. She makes a shed for a basic weft, and battens down the warp cross. The basic weft is not yet placed into the shed. The batten is left inside the shed, close to the edge of the weaving. The weaver now starts a row of orange extra wefts, inserting about one inch of each weft into the shed that is made by the batten. The first extra weft is inserted between warps 14 and 15 from the right, counting only the warps that pass over the batten. Twenty-eight warps farther to the left the next extra weft is put into the shed, and so on. The same procedure is repeated with the white extra wefts. The first white extra weft is inserted between warps 6 and 7. The following wefts are spaced evenly with 28 warps between two extra wefts. The warp stripes serve as counting aids. Inside the shed, orange and white extra wefts overlap, as shown in figure 65.

When all the extra wefts are positioned correctly, the weaver starts to make the first row of extra-weft floats. All the white extra wefts go to the right, over four, under one, over four, and under one warp. Warps that are needed to hold down an extra weft are lifted up with the fingers, and the extra weft is placed under these warps. The orange wefts are also placed under selected warps. White and orange wefts travel in opposite directions. The loose ends of the extra wefts remain on top of the fabric. Now the weaver sets the batten on edge, makes a pick of the basic weft and battens it down. She establishes the next basic shed and battens down the warp cross. This battening beats the extra wefts in place too. The batten remains in the shed, and the weaver starts the second row of extra weft floats. Compared to the first row of weft floats, the orange floats have now shifted one warp to the right, the white floats one warp to the left (see figure 66*a*).

Weaving Procedure

Step 1

Establish a basic shed, batten down the warp cross, leave the batten in the shed.

Step 2

Establish sheds for the extra wefts, and place the extra wefts in these sheds.

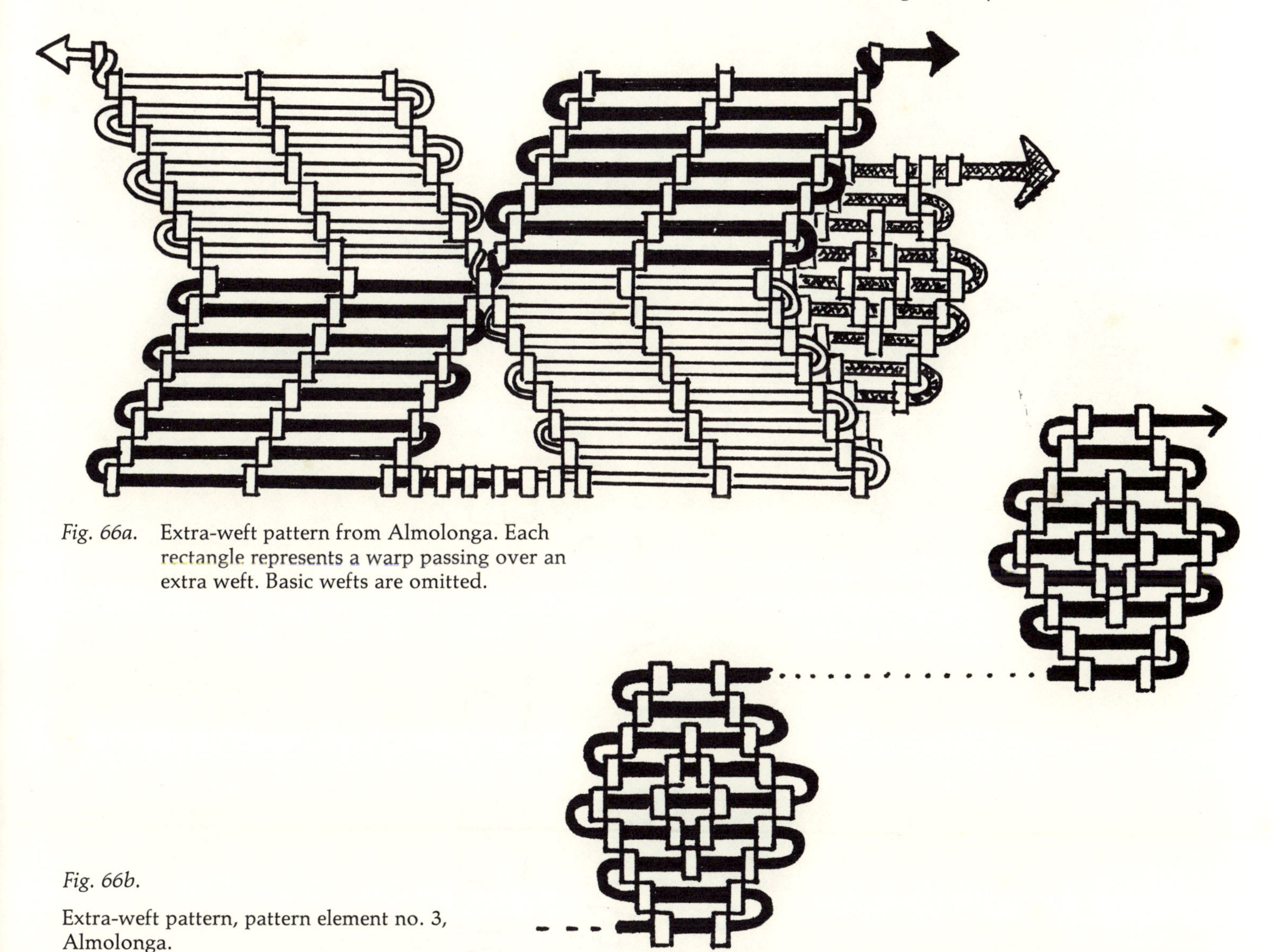

Fig. 66a. Extra-weft pattern from Almolonga. Each rectangle represents a warp passing over an extra weft. Basic wefts are omitted.

Fig. 66b.

Extra-weft pattern, pattern element no. 3, Almolonga.

Step 3

Put a basic weft into the shed that was held open by the batten. Batten down the weft.

Weaving the extra-weft pattern starts with pattern elements 1 and 2. After completing three rows of extra weft floats, the weaver starts 11 more weft floats to build up pattern element 3 (see figures 64 and 66*a*).

Figure 66*a* shows how white and orange extra wefts cross each other at certain points. Figure 66*b* shows two pattern element 3s separately. Between the two elements, the extra weft travels hidden in the basic shed.

Between two picks of the basic weft, the weaver has to arrange a row of 33 extra wefts. Extra wefts frequently come to an end and have to be started anew. Beginnings of new wefts are inserted into the basic shed. Weaving obviously proceeds slowly. We watched a woman weaving a huipil similar

Plates 55 and 56. Headband from Aguacatán.

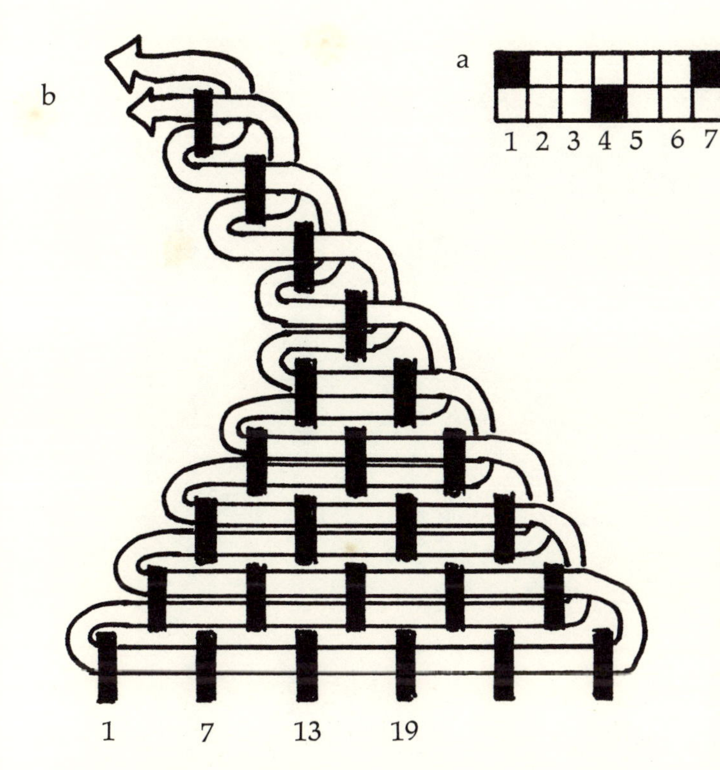

Fig. 67. Extra-weft pattern, Aguacatán: *a*, warps raised by *p*1 and *p*2; *b*, section of bird motif (foot).

to the one described here and came to the conclusion that it would take her at least 120 hours to do the brocading (5 minutes for every row of extra wefts). It must have taken more than 150 hours to weave the huipil that we describe here. The cost of the brocading yarn (Sedalina) comes to about twelve quetzales, the total cost of the huipil being fourteen quetzales. Only well-to-do women could afford such an expensive huipil.

HEADBAND FROM AGUACATAN

Special feature: Single-face brocading (plates 55 and 56).

Headbands are the most distinctive feature of the women's costumes in Aguacatán. Charming bird motives can be found on all headbands. Plant and abstract designs are also popular. Headbands are decorated their entire length. When the band is worn, only a small section of the pattern can be seen.

The headband shown in plate 55 is 2 yards and 16 inches long and 3½ inches wide. The ends are folded and sewn into triangular shapes. Three large multicolored tassels are attached to each end. The warp is three-ply red yarn. For the weft, three singles of an inexpensive red yarn are combined. The fabric has a warp surface, with 76 warps and 24 wefts per inch. Mercerized cotton yarn in blue, green, white, yellow, and purple is used as extra weft. Brocading is done with the aid of two pattern sticks shaped like small battens. Pattern stick *p*1 lifts up every third one of the odd-numbered warps, pattern stick *p*2 raises every third one of the even-numbered warps. Extra-weft patterns are of two kinds: *a*, horizontal bands produced by extra wefts that run across the whole width of the fabric and *b*, free-standing figures.

Weaving Procedure

Step 1

Establish a basic shed, batten down the warp cross, make a pick of the basic weft, and batten it down. The batten remains in the shed.

Step 2

With the aid of pattern stick *p*1 or *p*2 establish a shed for the extra wefts. Depending on the pattern motif, a continuous extra weft is placed into the shed across the whole width of the fabric, or else discontinuous wefts are placed into sections of the shed to form a free-standing figure. Go to step 1.

Plate 57. Section of a *servilleta* from Cotzal.

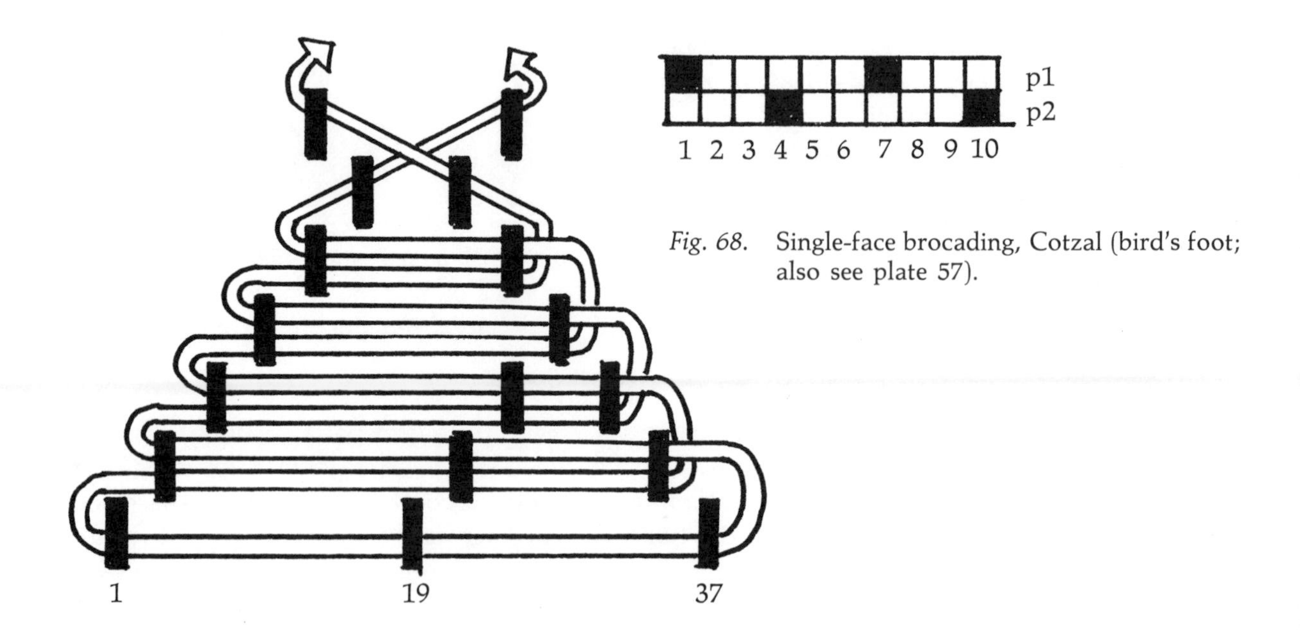

Fig. 68. Single-face brocading, Cotzal (bird's foot; also see plate 57).

Pattern stick *p*1 is used to make a brocading shed when the preceding basic weft is in shed 1. A basic weft in shed 2 is followed by extra wefts in a shed that is made by pattern stick *p*2. The manipulation of the pattern sticks has already been described in chapter 19, "Pattern Sticks for Single-Face Brocading."

Before starting on an extra-weft pattern, the weaver studies a sample of the motif that she wants to weave. She has to know which warps will be needed to hold down the extra wefts. With one of the pattern sticks she makes a shed that runs across the width of the fabric. Let us assume that she is about to start the bird pattern shown in plate 56. For each of the birds' feet she prepares a length of mercerized cotton. She counts the warps that pass over the pattern stick to find the exact place to start the first extra weft. The extra weft is placed under six consecutive warps so that the central section of the weft is under these warps. The ends of the extra weft are placed on top of the fabric. Each of the four extra wefts is put into place in this way. Thereafter the two ends of each extra weft always go into the same shed, traveling in opposite directions. Figure 67 shows a section of the bird motif. The extra-weft floats are all of equal length, passing over two odd- or even-numbered warps.

When an extra weft comes to an end and a new one is started, both wefts overlap in the same shed. When a figure is completed, the end of the extra weft is put into the next basic shed or left in the last brocading shed.

The floats of the extra wefts are short and the patterns are very precise, as can be seen in plate 56. Without the aid of pattern sticks, the weaver would have difficulty achieving similar results.

SERVILLETA FROM COTZAL

Special feature: Single-face brocading (plate 57).

Servilletas are used to cover baskets and to wrap tortillas and other foods. Weavers from Cotzal decorate their *servilletas* with birds and geometrical figures. Plate 57 shows a section of a *servilleta.* The *servilleta,* a panel 13½ inches by 24 inches, plus 2 inches of end fringes on two sides, has two selvages. The warps are two red single yarns combined, the warp stripes (blue, green, pink) are made with two-ply yarn. Four red singles are combined for the weft. The thread count is 66 warps and 19 wefts per inch. For the brocading several strands of two-ply yarn (blue, green, pink, white) are combined. The *servilleta* is decorated with two rows of birds and one row of geometrical figures. Brocading is done with the aid of pattern sticks (see chapter 19, "Pattern Sticks for Single-Face Brocading"). Two pattern sticks, *p*1 and *p*2, are used for the bird design, two more pattern sticks are used for the horizontal bands. Pattern stick *p*1 raises every third one of the odd-numbered warps, pattern stick *p*2 raises every third one of the even-numbered warps. Weavers in Aguacatán use a similar arrangement of pattern sticks. But the birds on headbands from Aguacatán look quite different from Cotzal birds, because the Cotzal weavers build up the bird figures with rather long weft floats. All the warps that hold down the weft floats are chosen from among the warps that are raised by *p*1 or *p*2. As in Aguacatán weaving, the two ends of the extra weft travel in opposite directions (see figure 68). The beaks of the birds are formed by twining the extra wefts around groups of warps.

HUIPIL FROM SAN ANTONIO AGUAS CALIENTES

Special features: Single-face brocading, double-face brocading, single-face wrapping, double-face wrapping (plates 58, 59, 60).

Weavers in San Antonio employ a number of techniques to decorate their huipils. They are very skillful in the use of pattern sticks. The huipil shown in plate 58 is a good example of fine San Antonio weaving. It is made from two panels, each panel has three selvages and one end that is cut and hemmed. The panels are sewn together at the sides, and holes are left for the head and arms. The huipil

Plate 58. One panel of a huipil from San Antonio Aguas Calientes.

is 21 inches long and 28 inches wide. The warp is a three-ply mauve yarn, and the same kind of yarn is used for the weft. The huipil has a warp surface, with 60 warps and 20 to 25 wefts per inch. Two-ply yarn and mercerized cotton in many colors are used as extra wefts. Different kinds of extra-weft patterns are arranged in horizontal rows as shown in plate 58. The same patterns are repeated on the back of the huipil. The uppermost pattern, along the shoulder, is done in a double-face technique. All the other patterns are single-face brocade.

Single-Face Brocading

Eight different kinds of single-face patterns are used to decorate the huipil. Some of them are rather complicated, combining continuous and discontinuous extra wefts. We shall discuss four of the simpler patterns.

Pattern 1

This pattern is characterized by horizontal bands in brickwork pattern. Rows of geometrical figures on the huipil are separated by narrow bands. There are ten such bands. The bands are produced by extra-weft floats that form a brickwork pattern. Two pattern sticks are employed to weave the bands. Stick *p*1 raises every fifth one of the odd-numbered warps (warps 1, 11, 21, and so on); stick *p*2 raises every fifth one of the even-numbered warps (warps 6, 16, 26, and so on). Every basic weft is followed by a pick of the extra weft. The extra wefts run across the whole width of the fabric.

Plate 60 shows four such bands.

Pattern 2

Another brickwork pattern also appears in plate 59, at the bottom. This pattern was woven with the same arrangement of pattern sticks as for pattern 1. However, unlike pattern 1, pattern 2 is woven with discontinuous extra wefts. The pattern consists of thirty-six geometrical figures in a row. Each figure is started with a different extra weft. The beginnings and ends of the extra wefts are hidden in the basic sheds. Two pattern sticks are employed to raise the warps that hold down the extra wefts. At certain points, extra wefts of two different colors pass under the same warp. This is done to avoid gaps in the pattern. Figure 70 shows a section of the pattern.

Pattern 3

The zigzag pattern in the top section of plate 59 is done with a special type of brocading resulting in a tapestrylike surface. The warps are completely covered by the extra wefts. On top of each basic weft are *two* picks of extra wefts. The extra wefts are entered into a shed and then, reversing their direction, into a countershed. This is shown in figure 71. We did not see how the weaving was done, but assume that two pattern sticks were used to raise the warps that pass over the extra wefts. One pattern stick lifts warps 1, 7, 13, and so on; the other stick lifts warps 2, 8, 14, and so on. The pattern is built up by discontinuous extra wefts. At no point do two different extra wefts pass around the same warp.

Pattern 4

The huipil is decorated along its lower edge with diagonal crisscross patterns (see plate 58 and *a* in figure 72). This pattern is produced by continuous extra wefts. In figure 72, *b* shows the basic pattern unit in block-diagram form. The black squares represent warps that hold down the extra wefts. The pattern requires four different brocading sheds (sheds A, B, C, D). Sheds A and B are established with the aid of pattern sticks *p*1 and *p*2. These two sticks remain in the loom throughout the weaving of the pattern. A third pattern stick, *p*3, is inserted into the warps to establish shed C. Pattern stick *p*3 interferes with the manipulation of stick *p*1. Whenever the weaver wants to use *p*1, she has to remove *p*3 from the loom. Since there are warp crosses between sheds C and D, it is not possible to keep a pattern stick in shed D. Every time shed D is needed, the weaver has to establish it anew by lifting up warps with a brocading sword. Pattern stick *p*3 remains in the loom while shed D is established.

Plates 59 and 60. Sections of a huipil from San Antonio Aguas Calientes.

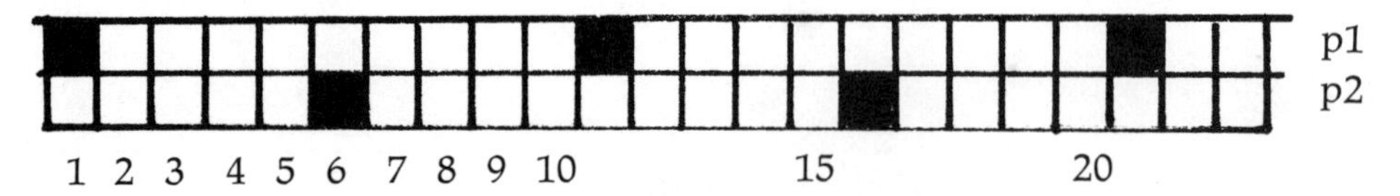

Fig. 69. Brickwork pattern, San Antonio Aguas Calientes.

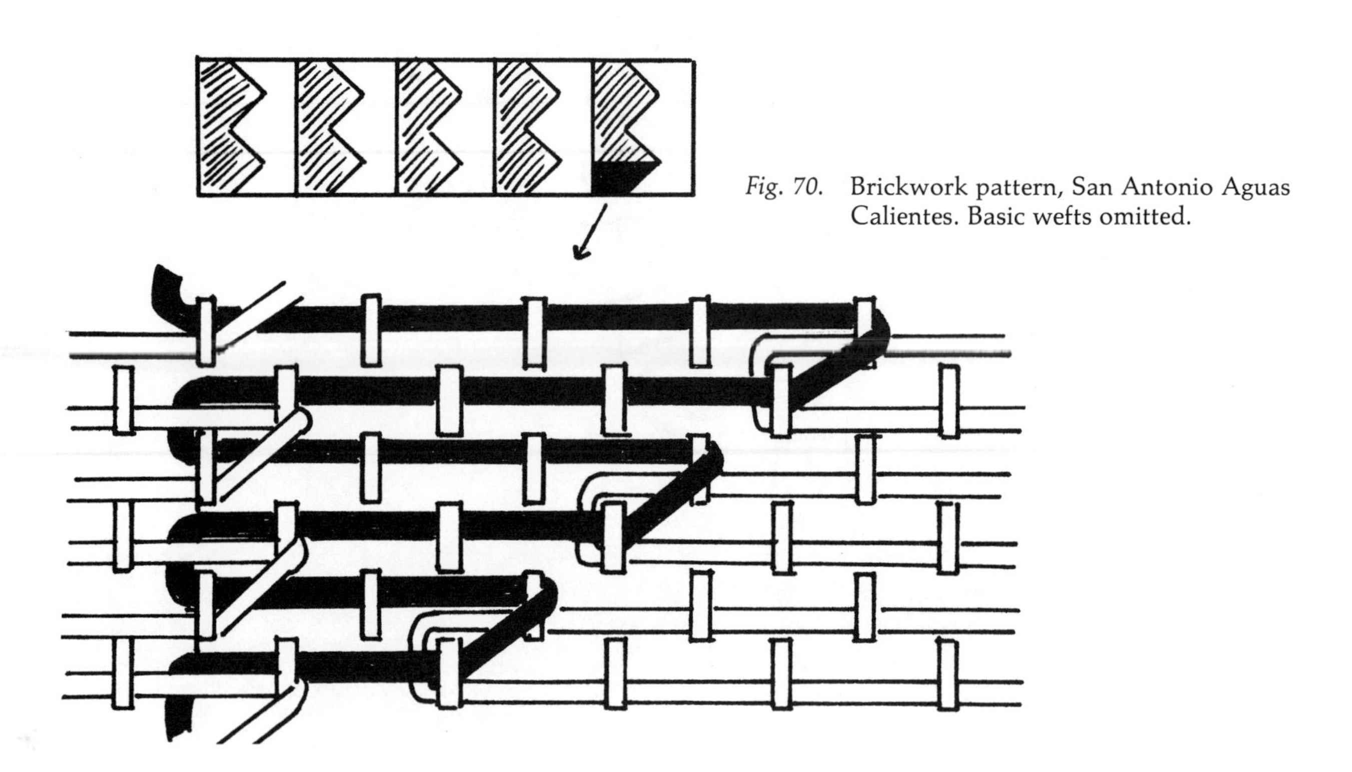

Fig. 70. Brickwork pattern, San Antonio Aguas Calientes. Basic wefts omitted.

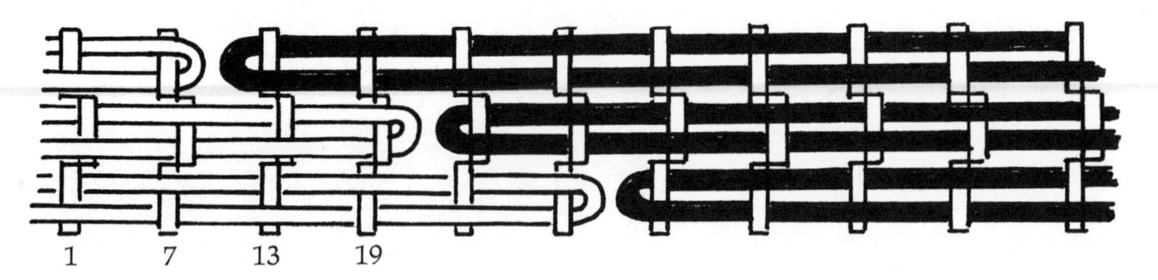

Fig. 71. Section of zigzag pattern, tapestry effect, San Antonio Aguas Calientes.

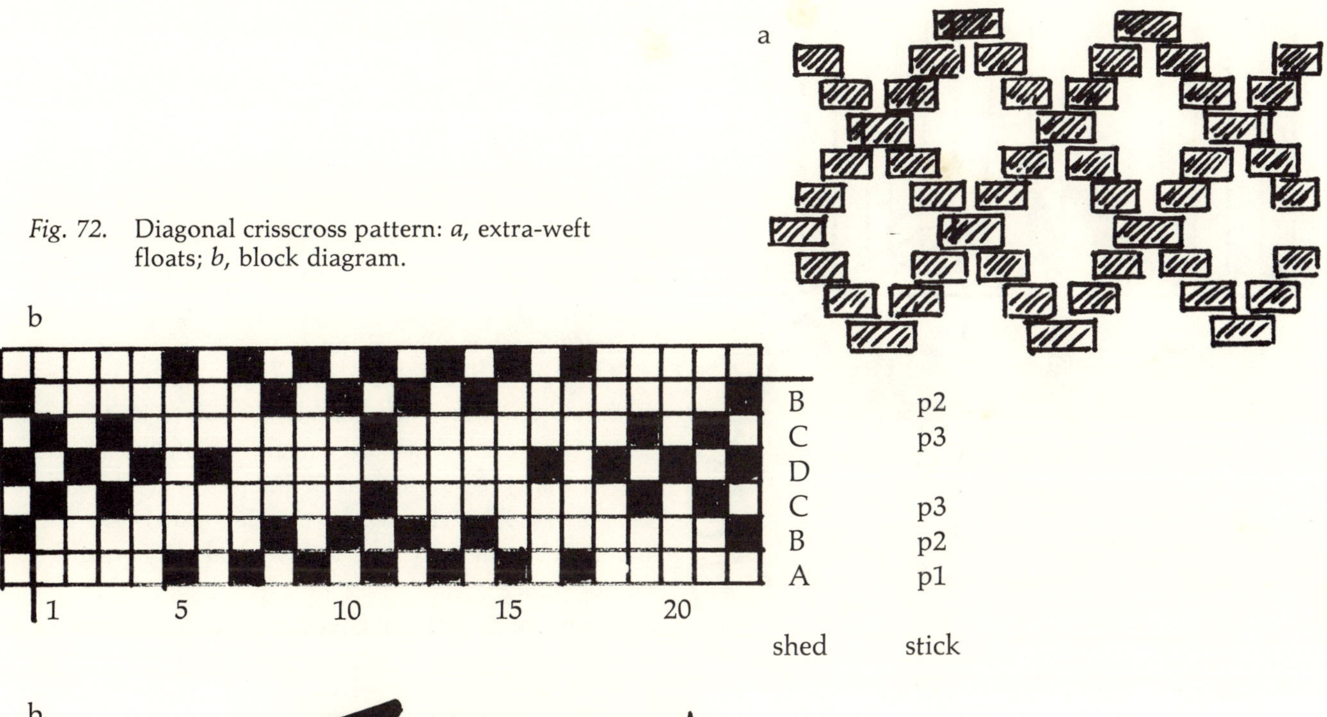

Fig. 72. Diagonal crisscross pattern: *a*, extra-weft floats; *b*, block diagram.

Fig. 73.

Double-Face Brocading and Wrapping

The floral pattern along the shoulder fold of the huipil is done in a technique combining double-face brocading and wrapping. This kind of weave seems to be of fairly recent origin in San Antonio. It is especially suited for copying designs from cross-stitch pattern books, since the extra-weft floats are of uniform length. These floats are arranged in vertical rows, as can be seen in plate 59 and figure 73.

Weaving Procedure

Each pick of the basic weft is followed by two picks of the extra weft. The extra wefts are entered into a shed and then, reversing their direction, into a countershed. The extra wefts pass over four and under four warps, as shown in figure 73. The resulting pattern appears on both sides of the fabric.

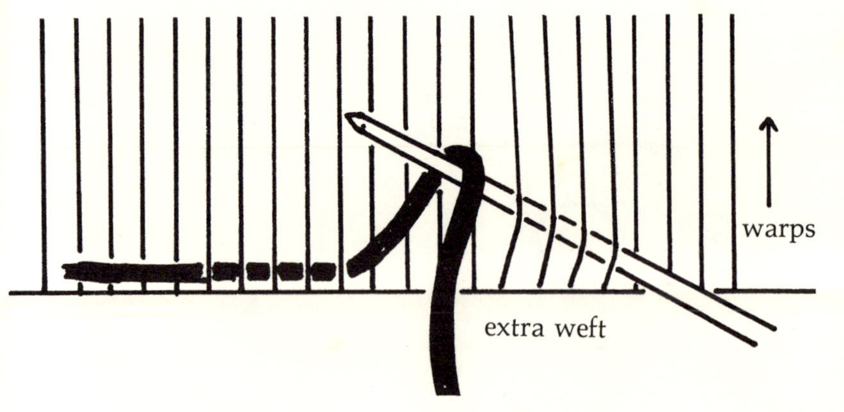

Fig. 74. Use of bone needle for double-face brocading, San Antonio Aguas Calientes.

Wefts of different colors are used for different parts of the pattern. When an extra-weft pick is finished, the loose end of the extra weft remains on top of the weaving.

In figure 73, *a* shows part of an embroidery pattern (flower petals), and *b* shows how this design is adapted to double-face brocading in the local style. Basic warps and wefts are indicated by thin black lines. The extra wefts appear as thick black lines where they are visible on top of the fabric and as dotted lines where they pass to the underside of the fabric.

The weaver uses a bone needle, about 7 inches long, to establish the sheds for the extra wefts. She lifts up four consecutive warps (odd- and even-numbered ones) with the needle, close to the edge of the weaving. The extra weft is looped around the tip of the needle (figure 74) and placed deftly into the shed. The bone needle is employed in a similar way for single-face brocading. It is also used to space warps when the loom is set up.

Building up a pattern from small extra-weft floats is a rather tedious task, as weavers pointed out to us. However, double-face patterns have become a trademark of San Antonio textiles. They are the main feature on textiles woven for tourists.

HUIPIL FROM NEBAJ

Special features: Weft surface, weft stripes, single-face brocading and wrapping, collar woven and embroidered separately (plates 61, 62).

Huipils in Nebaj are made from two or three panels. Of the three-panel huipil, there exists an elaborate version for ceremonial use and a simpler version that is for daily wear. The huipil shown in plate 61 is of the latter kind.

All three panels have three selvages and one hemmed end. The side panels are white and have a warp surface. They are machine stitched to the center panel. The huipil is 27 inches long. The center panel is 23 inches wide, the side panels are 9½ inches wide. In the following we deal only with the center panel.

A purple ribbon that was woven on the backstrap loom is sewn around the opening for the head. The ribbon is decorated with chain stitches. The center panel has a weft surface, and the warp count is low (about 14 warps per inch). Six white singles are combined for the warp. Different kinds of yarn are used for the weft. Where the huipil is decorated with diamond and chevron patterns, the weft consists of four red singles, and the weft count is about 26 wefts per inch. Two two-ply yarns are combined for the weft stripes. Groups of three or eleven weft stripes form horizontal bands. The stripes are green, blue, yellow, orange, red, white, and purple. The wefts are tightly battened down and the weft count is about 70 wefts per inch. For the lower section of the huipil and along the shoulder portion, four white singles are used as wefts. The extra wefts, consisting of four to eight strands of two-ply yarn or mercerized cotton are very thick. The colors are the same as for the weft stripes.

The center panel is decorated with rows of geometric figures and weft stripes. The figures are done in single-face brocading and wrapped weave (because of the low warp count, the extra wefts can also be seen on the underside of the weaving). Since the extra wefts are very strong, it is not

Plates 61 and 62. Huipil from Nebaj.

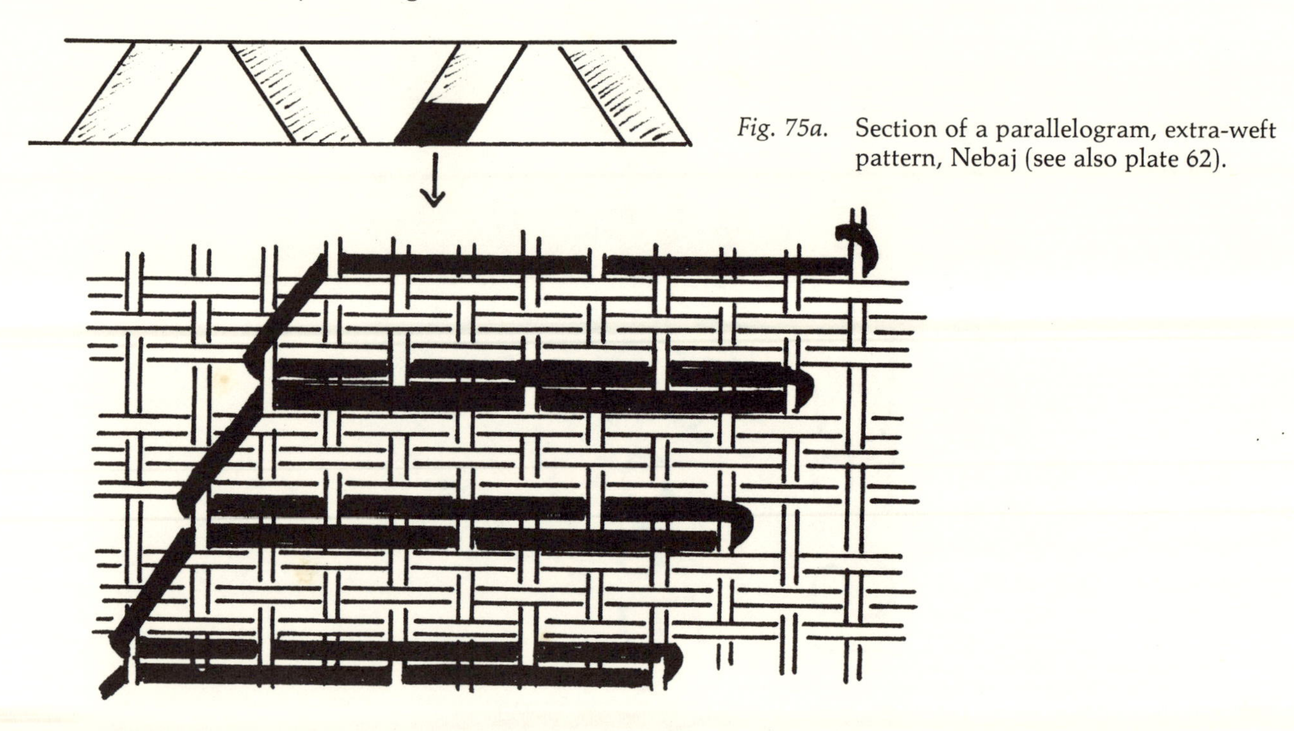

Fig. 75a. Section of a parallelogram, extra-weft pattern, Nebaj (see also plate 62).

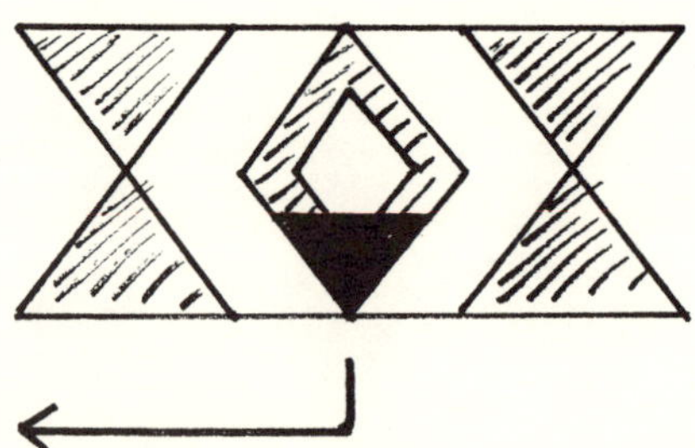

Fig. 75b. Section of a diamond-shape, extra-weft pattern, Nebaj.

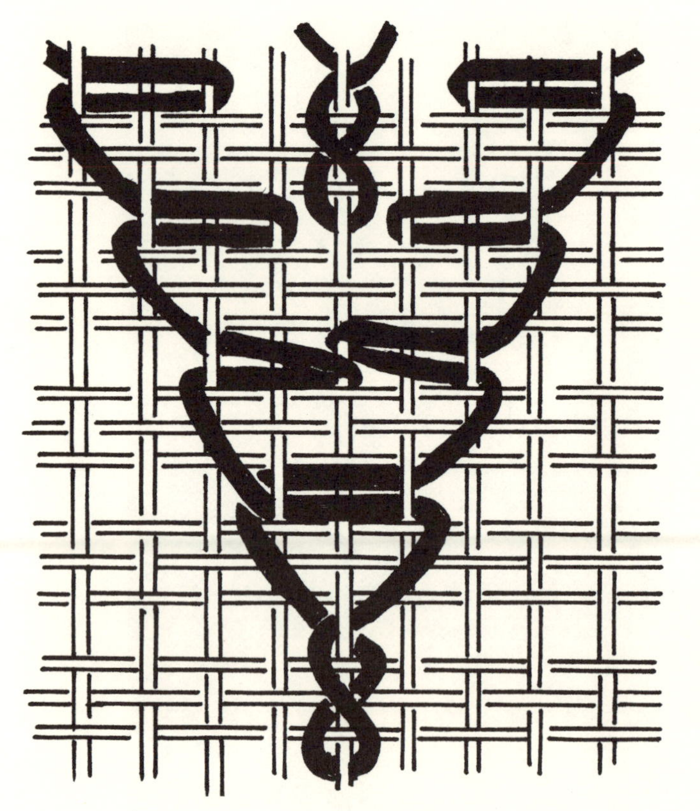

necessary to follow every basic weft with an extra weft. The sequence is three picks of the basic weft followed by two picks of extra wefts. In figure 75, two motives, a parallelogram, and a diamond shape are shown in detail. No special tools are employed to weave these patterns. Loose ends of extra wefts are placed into the basic shed.

Plates 63 and 64.
Huipil from San Martín Sacatepéquez.

HUIPIL FROM SAN MARTIN SACATEPEQUEZ

Special features: Two-face brocading, single-face brocading (plates 63, 64).

The huipil is woven in the traditional style, which nowadays is going out of fashion. Geometric patterns in dark blue are set against the red ground fabric. Here and there extra wefts of a lighter color serve as accents.

The two panels of the huipil were woven in one loom setup. Each panel has three selvages. The huipil is 26 inches long and 30 inches wide. It is open on the sides, except for a 3-inch-long section that is sewn together under the armholes. The warp consists of three red singles, the weft of six red singles. At the end selvages the wefts are extra strong. The thread count is 44 warps and 19 wefts per inch.

The extra wefts are mostly mercerized cotton yarns. Two-ply Mish is also used. The pattern is done in two-face brocade, except for four single-face horizontal bands.

Two-Face Brocading

The only extra tool needed is a brocading sword.

Weaving Sequence

Step 1: Pick of the Basic Weft.

A weft is put into a basic shed and battened down. The shed for the next pick of the basic weft is established. The warp cross is battened down. The batten remains in the shed, close to the heddle.

Step 2: Pick of the Extra Weft.

With the brocading sword, a shed is made for the extra weft. The extra weft is put into the shed. The batten, still in the basic shed (shed 1 or shed 2), is moved toward the edge of the weaving, thereby beating the extra weft in place. The batten is set on edge to open the shed for the next pick of the basic weft. Go to step 1.

Fig. 76. Two-face brocading, discontinuous extra weft, San Martín Sacatepéquez.

To establish the brocading shed, the weaver guides the brocading sword over or under selected groups of warps close to the edge of the weaving. No distinction is made between odd- and even-numbered warps. The brocading sword is set on edge, and the extra weft, which is wound on a bobbin, is passed through this shed, crossing the whole width of the fabric. With each pick of the extra weft, more than sixty weft floats are produced. The weaver has to make sure that the pattern develops properly. She has a sample at hand that serves as a model. The first brocading shed is done very carefully in order to get a symmetrical, evenly spaced pattern. The pattern is built up by weft floats of different lengths. The shortest float passes over two warps (one odd-, one even-numbered), the longest float goes over fourteen warps.

In addition to the dark-blue continuous extra wefts, discontinuous extra wefts of lighter colors are also employed. Wherever these wefts appear, the continuous extra weft is hidden in the basic shed. Figure 76 shows part of a discontinuous extra-weft pattern. The continuous extra weft is omitted. The ends of the discontinuous wefts are hidden in the basic sheds. After each pick, the discontinuous extra weft goes to the underside of the fabric (this is different from single-face brocading, where the extra weft stays on top of the fabric between two picks).

SERVILLETA FROM CONCEPCION CHIQUIRICHAPA

Special features: Two-face brocading, single-face brocading, end fringes (plate 65).

Nowadays, few weavers put much effort or imagination into weaving *servilletas.* The unusually well-made *servilleta* shown in plate 65 is an exception. Its main features are two rows of birds, one row at each end. Except for the section with the birds, the *servilleta* is decorated with horizontal ribs in single-face brocade. These ribs give more body and an interesting texture to the cloth. Two ends of the *servilleta* are nicely finished with knotted end fringes.

The *servilleta* is 15 inches wide and 29 inches long, including 3 inches of end fringes at both ends. The cloth has two selvages. Red cotton yarn is used for warp and weft. Warps consist of three singles, wefts of four singles. There are 46 warps and 20 wefts per inch. For the birds, dark-blue two-ply yarn is used. The single-face brocading (horizontal ribs) is done with white two-ply yarn and red single yarn (eight singles combined).

Two-Face Brocading

The bird motives are woven with continuous extra wefts, and the weaving technique is similar to the one used for the huipil from San Martín Sacatepéquez. The pattern is different from the one on the huipil from San Martín, but there are some similarities. Diagonals and, to a lesser extent, vertical lines are emphasized. With every new pick of the extra weft, the weft floats are shifted a fixed number of warps over to the right or the left, producing the diagonals.

Plate 65. Section of a *servilleta* from Concepción Chiquirichapa.

Single-Face Brocading

Directly above and below the bird motives appear six rows of white extra-weft floats. The extra wefts are put loosely into their sheds to give more body and texture to the *servilleta.* The largest part of the *servilleta* is decorated with red extra-weft floats. These floats are the same color as the basic warps and wefts, and the effect is mainly a textural one, as can be seen in plate 65. The extra-weft floats are woven with the aid of an extra heddle and a pattern stick. The weaving procedure has already been described in chapter 19, "Looms Equipped for Single-Face Brocading."

Plates 66 and 67. Huipil from Concepción Chiquirichapa.

HUIPIL FROM CONCEPCION CHIQUIRICHAPA, ALDEA TUIPOX

Special features: Two-face brocading with interlocking extra wefts (plates 66, 67).

This kind of huipil is popular in Concepción Chiquirichapa and San Martín Sacatepéquez. Foot-loom weavers in the Quezaltenango-Totonicapán area use similar brocading patterns. The upper section of the huipil is decorated with rows of animals, birds, flowers, and geometrical motives. The lower part of the huipil is worn under the skirt. Starting at the shoulder fold of the huipil, in row one we find geometric shapes, in row two owls and flowers, in row three squirrels and flowers, in row four cats and flowers, and in row five frogs and little birds. These motives are repeated in the following five rows. Row eleven, a checkerboard pattern, is also done in two-face brocading.

Both panels of the huipil were woven in one loom setup. Each panel has three selvages. The huipil is 26½ inches long and 28 inches wide. A rectangular hole has been cut for the head. The armholes and the opening for the head are finished with black velvet ribbon. Below the armholes, the front and back parts of the huipil are sewn together for about 3 inches. Farther down, the huipil is left open at the sides. Warps and wefts are red. Three singles are combined for the warp, four singles are used as wefts. The thread count is 40 warps and 24 wefts per inch. For the extra wefts the weaver combined several strands of two-ply yarn.

Except for the geometric shapes in the top row and the checkerboard pattern, all figures are woven with discontinuous extra wefts. On the right side of the huipil the different figures in one row are clearly separated from each other. On the underside of the fabric, however, extra wefts of adjacent figures interlock with each other, as shown in figure 77.

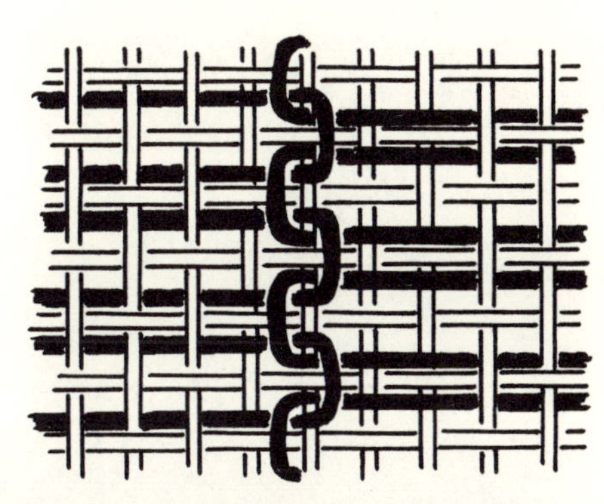

Fig. 77. Interlocking extra wefts on the wrong side of the fabric. The basic warps and wefts are white, the extra wefts black.

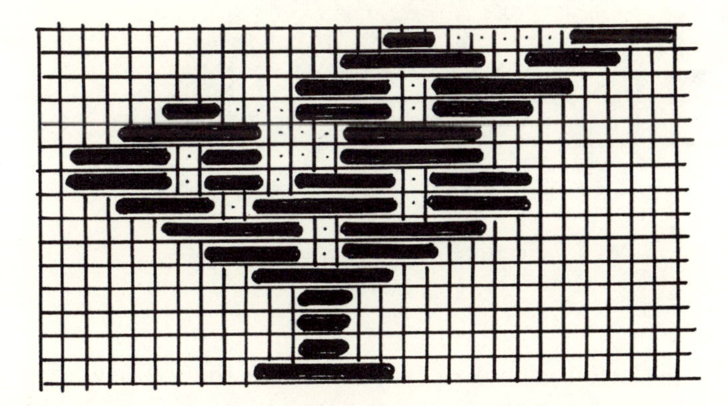

Fig. 78. Two-face, extra-weft pattern on huipil from Concepción Chiquirichapa. The thin black lines represent basic warps and wefts. The heavy black lines represent extra-weft floats. Dots indicate that the extra weft is on the underside of the fabric. The interlocking of the extra wefts is *not* shown.

Weaving Procedure

The weaver prepares the required number of extra wefts for starting the pattern. She makes a knot at the end of each extra weft. A shed is established for the next basic weft, and the batten is left in this shed, in flat position. Close to the edge of the weaving and using her fingers, the weaver separates the warps at regular intervals and inserts extra wefts into these gaps so that the knots are positioned under the warps. Starting at the right-hand side, the weaver produces the first row of weft floats by passing each extra weft over and under

Plate 68. Section of a huipil from Nahualá.

selected groups of warps. Where the extra warps are to pass from the upper side to the underside (or vice versa), the weaver makes a gap between the warps with her fingers. When the last weft float for one figure is done, the extra weft is left dangling on the underside of the fabric. After a row of weft floats has been completed, the batten is set on edge to open the shed for the basic weft. A basic weft is put into the shed and battened down and the extra wefts are battened into position at the same time. The next basic shed is made and the batten is put into the shed. The weaver then starts the second row of extra-weft floats from the left-hand side. The extra wefts now travel from left to right. Whenever the weaver picks up a new extra weft, she loops it around the preceding extra weft on the underside of the fabric and then brings it to the surface. This procedure connects the different extra wefts on the underside of the fabric and prevents warps from being pulled out of place by extra wefts.

HUIPIL FROM NAHUALA

Special feature: Two-face brocading (plates 43 and 68).

Many textiles from Nahualá are decorated with complex patterns done in very fine two-face brocading. Because of the great variety of motives—birds, lions, horses, and many other animals—Nahualá weavings are among the most interesting textiles from Guatemala.

Unlike the huipils discussed so far, huipils from Nahualá do not have the same pattern on the front and back. Our specimen is decorated with birds, bats, lions, dogs, a cat, and a row of geometrical ornaments. The largest bird motif and part of the geometrical pattern is shown in plate 68.

Two panels are sewn together to form the huipil. The garment is 38 inches long and 40½ inches wide. Holes are left for the head and arms. Each panel has four selvages and a join, which appears about 15 inches above the lower edge of the huipil. Weavers in Nahualá make rather long heading strips and put little effort into weaving the join, which, after all, cannot be seen when the lower part of the huipil is tucked into the skirt. Two singles of white *hilo chino* (bleached yarn) are combined for warps and wefts. There are 60 warps and 25 wefts per inch. Because of the fine yarn (size no. 16), weaving is done with more caution than in other places where stronger yarn is used. Most of the brocading is done with a purple-colored artificial silk (rayon). Two-ply yarn in many colors is also used. The rayon yarn stains when wet, producing a crushed raspberry color. This effect is purposely produced by soaking the newly woven huipil overnight in soapy water.

The pattern at the shoulder fold stretches across the whole width of the huipil and is, to a large part, woven with continuous extra wefts. When weaving the larger figures, the weaver uses a brocading sword to establish the sheds for the extra wefts (see plate 43). The pattern is built up by short floats that pass over two, four, or six consecutive warps. The rayon yarn used for the brocading is difficult to handle because it tends to curl. Six plies

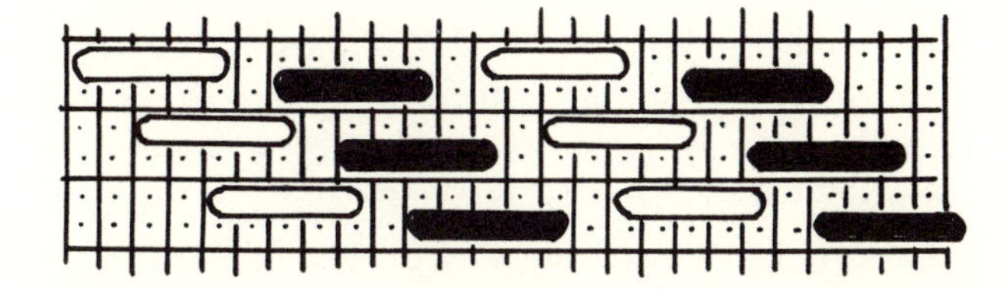

Fig. 79. Two-face brocading with two extra wefts.

of rayon are combined for an extra weft and knotted together at one end. The other end is inserted between the warps at the place where the brocading is to start. The extra wefts that consist of two-ply yarns are knotted at both ends to prevent them from slipping out between the warps. The weaving sequence—basic wefts alternating with rows of extra wefts—is the same as described in previous chapters. For some parts of the pattern, such as the tail and wing of the bird shown in plate 68, *two* extra wefts of different colors are put in between picks of the basic weft. These extra wefts travel in opposite directions, each one in its own shed. This is shown in figure 79.

The bird in plate 68 is woven with extra wefts of eight different colors. Only short floats of the extra wefts are visible on the right side of the weaving. The underside, however, looks quite different: long floats and extra wefts of different colors cross each other, bridging the gaps between two elements of the same color.

HUIPIL FROM CHICHE

Special features: Two-face brocading, pile loop surface.

A few decades ago women in Chiché and Chichicastenango decorated their huipils with such traditional motives as the double eagle. Nowadays huipils in the old style are collector's items. A typical modern huipil looks like the one that is worn by the Chiché weaver shown in the color plates (see page 70). The pattern is probably copied from an embroidery pattern book; the texture reminds one of a rug. Short weft floats that form little loops are arranged in vertical rows.

Unfortunately, our only huipil from Chiché was stolen, and we do not have the measurements of the garment. Like all huipils from Chiché and Chichicastenango, it was made from three panels, each panel having four selvages. The panels are joined together by a *randa,* a decorative band of embroidery stitches. In the center panel a hole is cut for the head, and a collar with embroidered flowers is sewn around the opening. Such collars are made by specialists and are sold on the market. The center panel of the huipil is brocaded down to the waist. The side panels are brocaded in the upper section.

In the past, unbleached white or natural brown cotton yarn was used for warps and wefts. Today weavers prefer two-ply yarn for the warps and several singles combined for the weft. The warp count is about 60 to 70 warps per inch.

Most weavers use two-ply yarn or acrylic wool in gaudy colors for the extra wefts. We were told that one and a half pounds of two-ply yarn is needed for the brocading, and the same amount of yarn is necessary for the ground fabric.

Two-Face Brocading

Weavers in Chiché and Chichicastenango prefer to use patterns from cross-stitch pattern books. Occasionally single sheets with cross-stitch patterns are sold on the market. Such patterns are built up by numerous small squares. To produce a similar effect, the backstrap weaver has to make small floats of a uniform length. Two such floats running parallel with each other are needed to reproduce the basic unit of a cross-stitch pattern. Weavers employ two pattern sticks to establish the shed for the extra wefts. The shed consists of two warps up, six warps down, and extends over the whole width of the fabric except for a small margin at the edges. Every extra-weft float goes over six warps. Every basic weft is followed by a row of extra wefts. The procedure for making the shed for the extra wefts has already been described in chapter 19, "Pattern Sticks for Two-Face Brocading."

Most huipil patterns feature multicolored flowers against a solid background. Such patterns are built up by many different extra wefts. If burgundy is the background color, the first section of a brocading shed is filled with a burgundy-colored extra weft. A few floats of green weft follow to build up the leaves. Then comes a red weft for the rose petal, and so forth. No section of the brocading shed is left empty.

Weaving Procedure

To start an extra weft, the weaver prepares a length of brocading yarn (four strands of two-ply yarn combined) and makes a knot at one end of the extra weft. The knotted end is inserted between the warps at a point where the pattern requires it. The knot keeps the weft from slipping out from between the warps.

When entering the extra weft into the brocading shed, the weaver gives it a slanting position (plate 46). Each float is pulled with a hooking motion of the index finger towards the edge of the weaving and beyond, so that it forms a loop. These loops are characteristic of Chiché weaving.

After the last float, the extra weft goes to the underside of the weaving. It reemerges when the next row of weft floats is made, if it is not already used further down in the same row. Establishing the brocading shed, counting the floats, starting new extra wefts, and keeping track of different figures being built up at the same time are tasks that require patience and concentration. Every single weft float must be pulled outwards with the finger to form a loop. Sometimes loops are too long or too short and have to be corrected with a needle. There are more than 100,000 such loops on a single huipil. After watching a weaver for several hours, we came to the conclusion that weaving three panels for a huipil would take her 150 to 180 hours.

HUIPIL FROM ZACUALPA

Special features: Twining, two-face wrapped weave, warp stripes (plates 69, 70).

Wrapping and twining are methods of thread interlacing that were already in use before weaving was invented. Weavers in Zacualpa use these techniques mainly to add decorative elements to their textiles. Compared to most brocading techniques, wrapping and twining are very slow procedures. This may be one reason why only small sections of the textiles are decorated in this way.

The huipil is 33 inches long and 29 inches wide. Two panels, each having three selvages, are joined with large sewing stitches. Openings are left for the head and arms. Below the opening for the head, the two panels are sewn together with decorative stitches that form a *randa.* Two-ply yarn is used for warps and wefts. Groups of about fourteen red warps are separated by two orange warps. About every two inches there is a green and yellow warp stripe (six green, two yellow, six green warps).

More complex stripes, with odd- and even-numbered warps of different colors, appear close to the side selvages. The weft yarn is red. The huipil has a warp surface, with about 90 warps and 20 wefts per inch.

The back and front of the huipil are decorated with similar extra-weft patterns. One inch below the shoulder fold is a horizontal band of twined wefts. Above and below this band, dark purple extra wefts are wrapped around groups of warps. The twined as well as the wrapped wefts form short slanting floats that are raised above the adjacent areas of plain weaving.

Wrapped Weave

Plate 70 shows a section of the wrapped weave pattern. It is formed by extra wefts that travel on a zigzag path. Since the extra wefts are placed close together, they cover the warps of the ground fabric. The extra wefts that appear dark in the picture are of a dark-purple color (mercerized cotton). A few extra wefts are green, blue, or orange (two strands of two-ply yarn combined). In figure 80, *a* shows how the extra wefts are started. The weaver needs about ninety extra wefts to start the pattern. Each extra weft is 16 inches long. The warps are divided into groups of six, and an extra weft is inserted under every other group, as shown in *a* of figure 80. The ends of the extra wefts, each one

Plates 69 and 70. Huipil from Zacualpa.

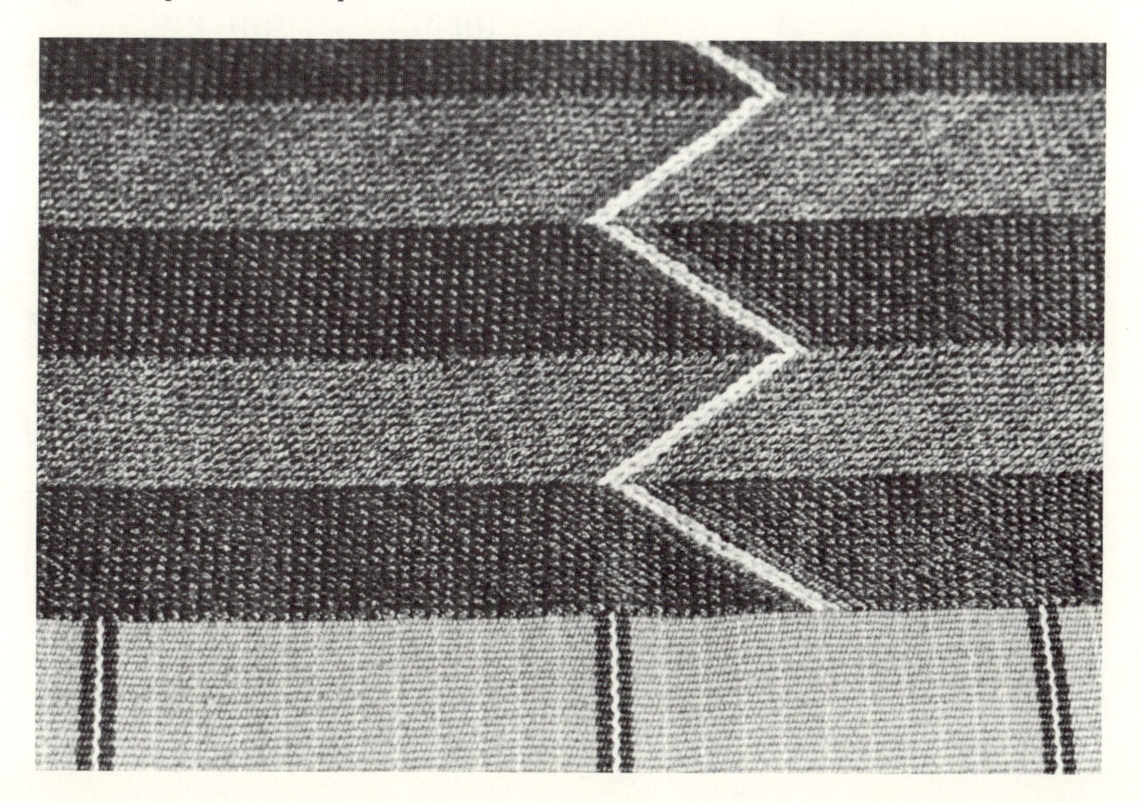

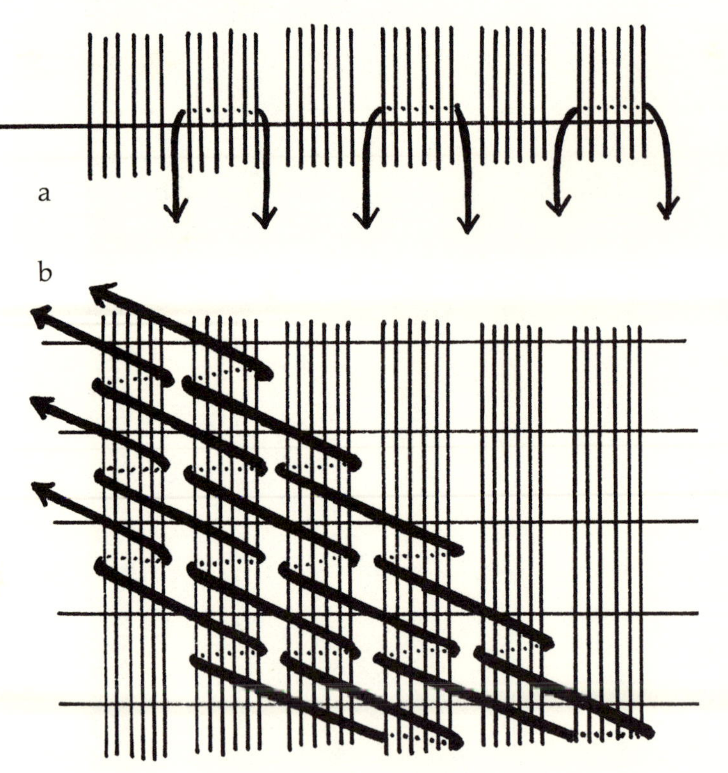

Fig. 80. Wrapped weave, Zacualpa: *a*, starting the extra weft; *b*, extra-weft pattern.

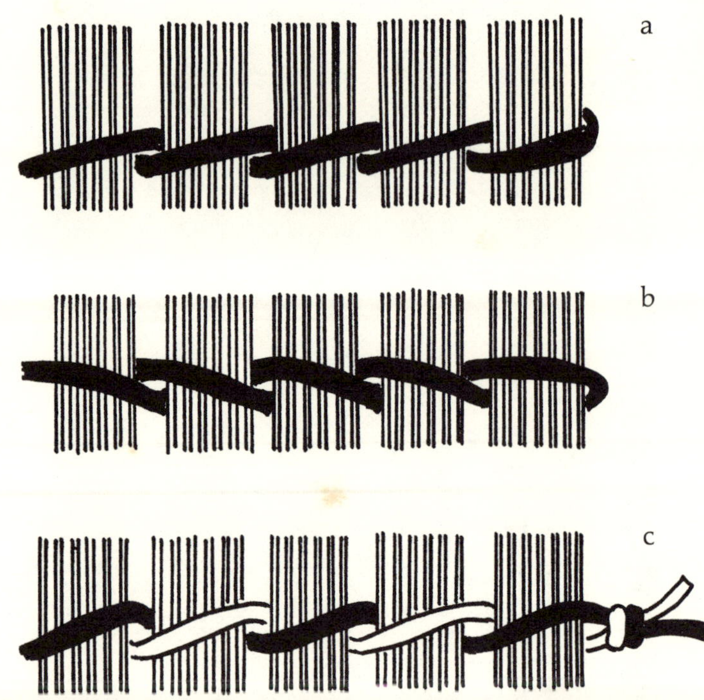

Fig. 81. Twining, Zacualpa.

about 8 inches long, are on top of the fabric. When all the extra wefts are in place, the weaver makes a pick of the basic weft. The basic weft is battened down and the next basic shed is established. The warp cross is battened down, and the batten remains in the shed, some distance from the edge of the weaving. Now, each end of an extra weft is wrapped around a group of six warps, as shown in *b*, figure 80. When all the extra wefts have been wrapped around the corresponding warp groups, the batten is set on edge, and a basic weft is put into the shed. From now on picks of the basic weft alternate with manipulation of the extra wefts. For about fifteen picks, the extra wefts travel to the left. For the next fifteen picks they move towards

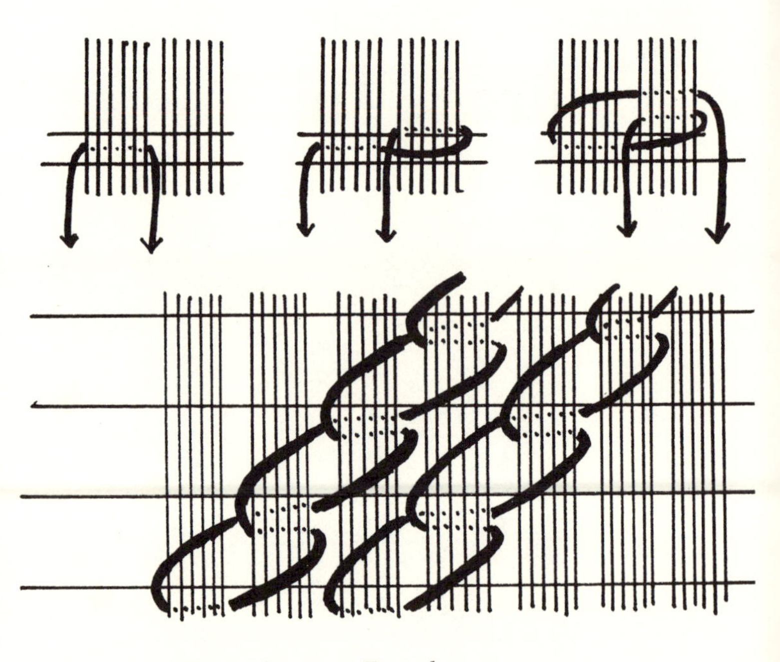

Fig. 82. Wrapped weave, Zacualpa.

the right, and so on. In plate 70, the extra wefts appear darker where they travel to the left and they reflect more light when traveling to the right. This effect is caused by the twist of the yarn. Weft floats appear darker or lighter depending on the angles between the light source, the yarn fibers, and the viewer's eye. Weavers are aware of this effect and emphasize it by using purple wefts with a shiny surface.

It takes about ten minutes to wrap one row of extra wefts around the corresponding warps. Since each panel of a huipil has approximately two hundred rows of wrapped extra weft floats, it takes about seventy hours to complete the wrapped weave pattern for an entire huipil. After the last row of floats, the ends of the extra wefts are brought to the underside of the weaving where they form fringes.

Twining

Close to the shoulder fold, breaking up the wrapped-weave area, are two horizontal bands of twined wefts. The wefts consist of four two-ply yarns combined. They are twined around groups of six warps, the same groups that were used for the wrapped weave.

After traveling across the whole width of the panel, the two weft ends are left hanging down at the side. Some weavers braid the loose ends. The finished row of twined wefts is pushed down toward the working edge of the weaving with the fingers. Altogether there are ten rows of twined wefts in one band. Two of these rows are done in two colors, as shown in *c*, figure 81.

The twined wefts are not superimposed over the basic wefts, they replace them. Removing the twined wefts would leave a gap in the fabric, with only the bare warps showing.

Other Kinds of Wrapped Weave

Zacualpa weavers know several methods of wrapping. In figure 82, we illustrate a pattern found on another huipil.

HUIPIL FROM COLOTENANGO, ALDEA GRANADILLO

Special features: Single-face brocading, single-face wrapped weave, warp stripes (plates 71, 72).

This huipil represents fine Colotenango weaving in the traditional style. Warp stripes and extra-weft patterns are harmoniously arranged and the color combination—red and white with a few accents in green, yellow, and orange—is very successful. The huipil is 20 inches long and 42 inches wide. The center panel is 16 inches wide and the side panels are 15 inches and 15½ inches wide. The center panel overlaps the side panels by about 2 inches on each side. Each panel has four selvages. The joins are carefully woven and can hardly be noticed. The openings for the head and arms are hemmed with decorative stitching in green and yellow.

Three kinds of yarn are used for the warps: hand-twisted white singles, hand-twisted red singles, and paired orange singles. The warp stripes are evenly spaced over the whole width of the panels. The warps are arranged in the following sequence: 6 red, 2 orange, 16 red, 2 orange, 6 red, 54 white, 16 red, and 54 white warps. This sequence is repeated eight times, and a red and orange stripe is added, bringing the total to 1,280 warps in each panel. Warps of different colors are knotted together.

For the wefts, single white yarn is used. The end selvages are woven with paired singles. The fabric has a warp surface, with 82 warps and 29 wefts per inch.

Extra-Weft Patterns

Most of the extra-weft patterns are done with single red yarn. Four singles are loosely twisted together to make a heavier extra weft. Mercerized cotton in green, yellow, and purple is used for some of the patterns.

All patterns are done in single-face technique. No special tools are employed.

Plates 71 and 72. Huipil from Colotenango.

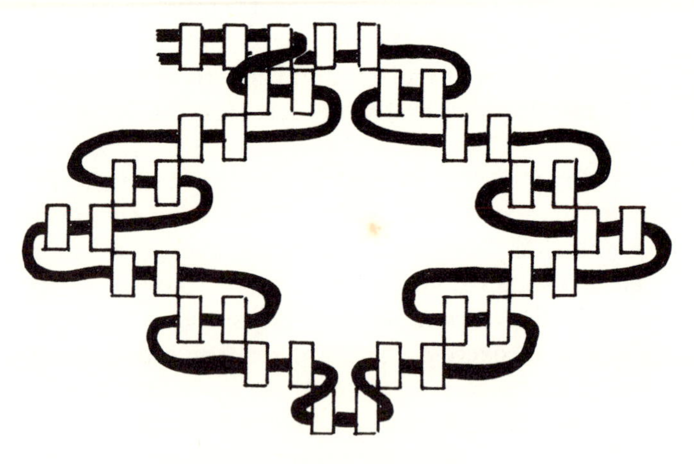

Fig. 83. Rosette pattern, Colotenango.

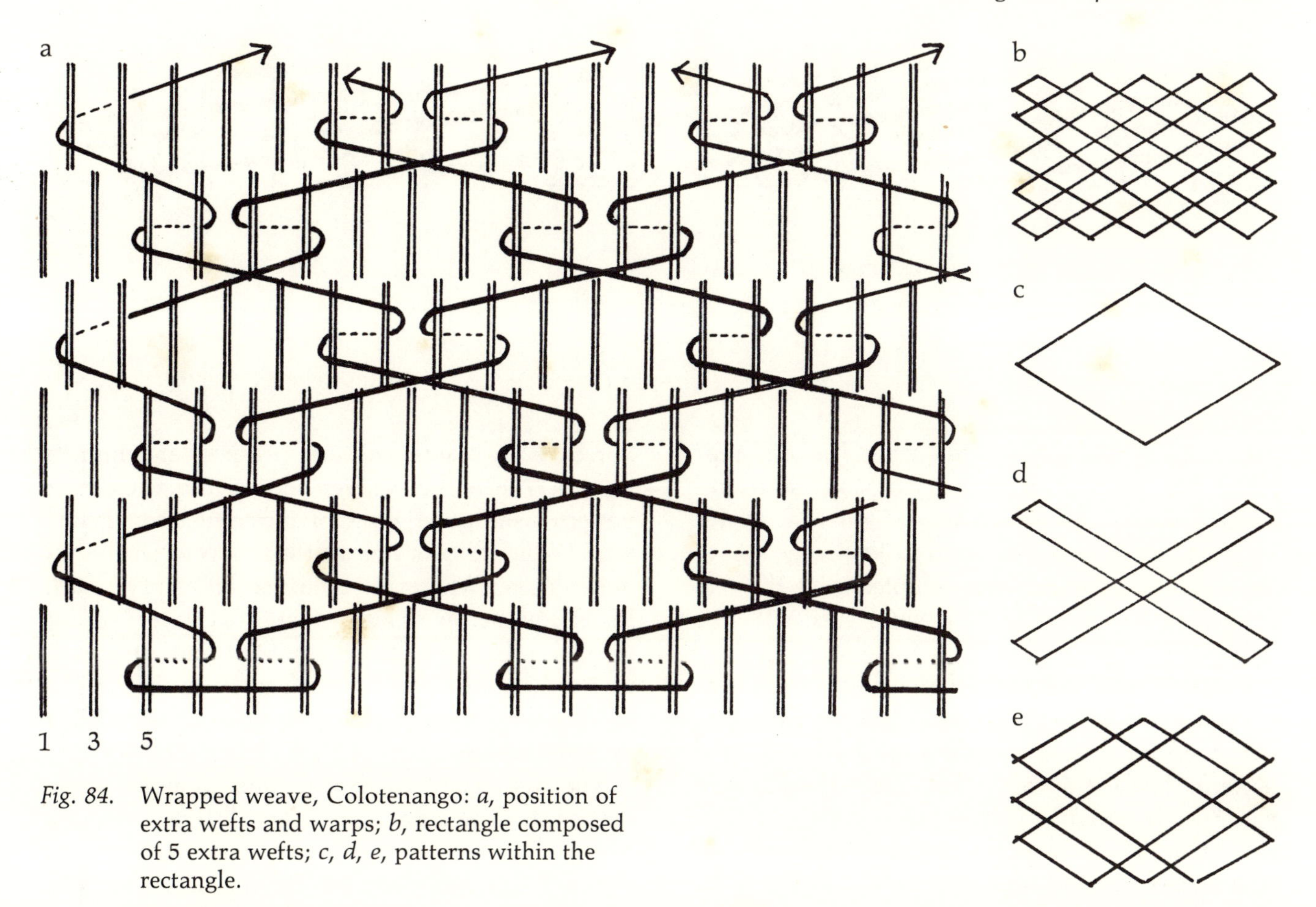

Fig. 84. Wrapped weave, Colotenango: *a*, position of extra wefts and warps; *b*, rectangle composed of 5 extra wefts; *c*, *d*, *e*, patterns within the rectangle.

Weaving Sequence

Step 1

Establish a shed for the next basic weft. Batten down the warp cross. Put the basic weft into the shed. Leave the batten in the shed and set it on edge.

Step 2

Make sheds for the extra wefts (or wrap the extra wefts around selected warps), and put the extra wefts into these sheds. Go to step 1.

Every basic weft is followed by a row of extra weft floats.

Three kinds of pattern elements can be distinguished on the huipil: rosettes, rectangles, and horizontal bands. All are raised above the ground fabric, and it is mainly the textural effect that makes the pattern interesting (plate 72).

Each one of the rosettes—there are more than 1,200 of them on the huipil—requires an extra weft. The pattern is built up in nine consecutive steps, with the two ends of the extra weft traveling in opposite directions (see figure 83). The weaver uses her fingers to lift up the warps that are to pass over the extra wefts. Two warps are lifted up together. When the pattern is finished, the ends of the extra wefts are put into the basic shed. Because of their thickness, they form a ridge, or weft cord. This textural effect is intended by the weaver.

The rectangles are placed between the warp stripes, sixteen rectangles in one row. Each rectangle is formed by several extra wefts. Again, the

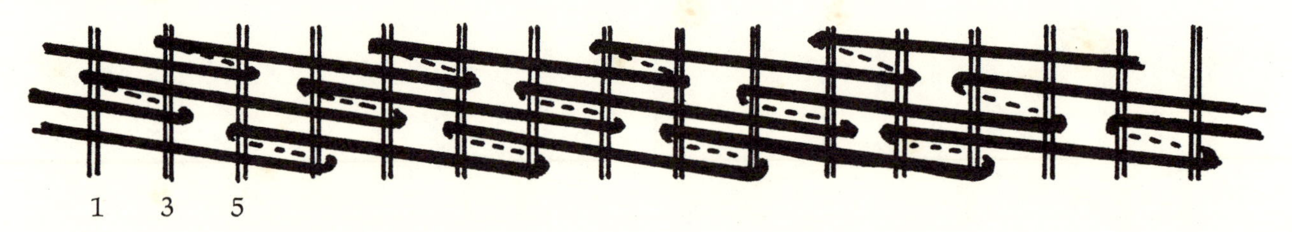

Fig. 85. Wrapped weave, Colotenango.

two ends of the extra wefts travel on different paths. The weaver wraps the extra wefts around selected warps, as shown in *a*, figure 84. Plates 40 and 41 show a woman from Colotenango weaving a similar pattern. After completion of the rectangle, the weft ends are put into a shed with the basic weft.

By using extra wefts of different colors the weaver produces diamond patterns within the rectangle. Every second rectangle on the huipil has a diamond pattern. The extra wefts that form the diamond patterns are green or yellow.

In figure 84, *b* shows a rectangle composed of five extra wefts. Sections *c*, *d*, and *e* of figure 84 show three patterns that can be made within the rectangle.

Horizontal lines—two are shown in plate 72—are also done in a wrapping technique. Each band is formed by three extra wefts, two red and one yellow. Figure 85 shows how this pattern was established.

HUIPIL FROM NEBAJ

Special features: Single-face wrapped weave, single-face brocading (plate 73, 74).

Because of their bold, colorful designs, textiles from Nebaj are popular with tourists. Weavers have increased their production, and the quality of the weaving has declined. A well-made huipil like the one in plate 73 is a rarity. The huipil is decorated with such typical Nebaj motives as birds, horses, corn plants, flowers in a pot, and human figures. When the weaver started the extra-weft patterns, she put figures of different sizes in one row. While still working on the first row, she began new motives wherever she found an empty space. In general, her experimental approach worked out well, and the different figures do not interfere with each other. At some points, however, she had to modify the standard motives to make them fit the available space.

The huipil is 27 inches long and 48 inches wide. It consists of two panels, each with three selvages. The panels are sewn together on a sewing machine. A round opening is cut out for the head. Sections of handwoven purple ribbon are sewn around the opening and decorated with chain stitches.

The white ground fabric is very strong and firm. Three singles are combined for the warp, four singles go into the weft. The thread count is 70 warps and 18 wefts per inch.

For the extra wefts two-ply yarn, or mercerized cotton, was used. Each figure is woven with several extra wefts of different colors. The weaving sequence (for the extra weft patterns) is the same as that described for the huipil from Colotenango. Each basic weft is followed by a row of extra wefts. All patterns are single-face, and no such tools as pattern sticks or brocading battens are used.

Women in Nebaj employ several methods of wrapped weaving. Figure 86 illustrates five of them. In plate 74 we can see how these methods are combined to build up the different pattern elements. Large parts of the design are done in a figure-eight wrapping technique (figure 86, *c*), which is char-

Plates 73 and 74. Huipil from Nebaj.

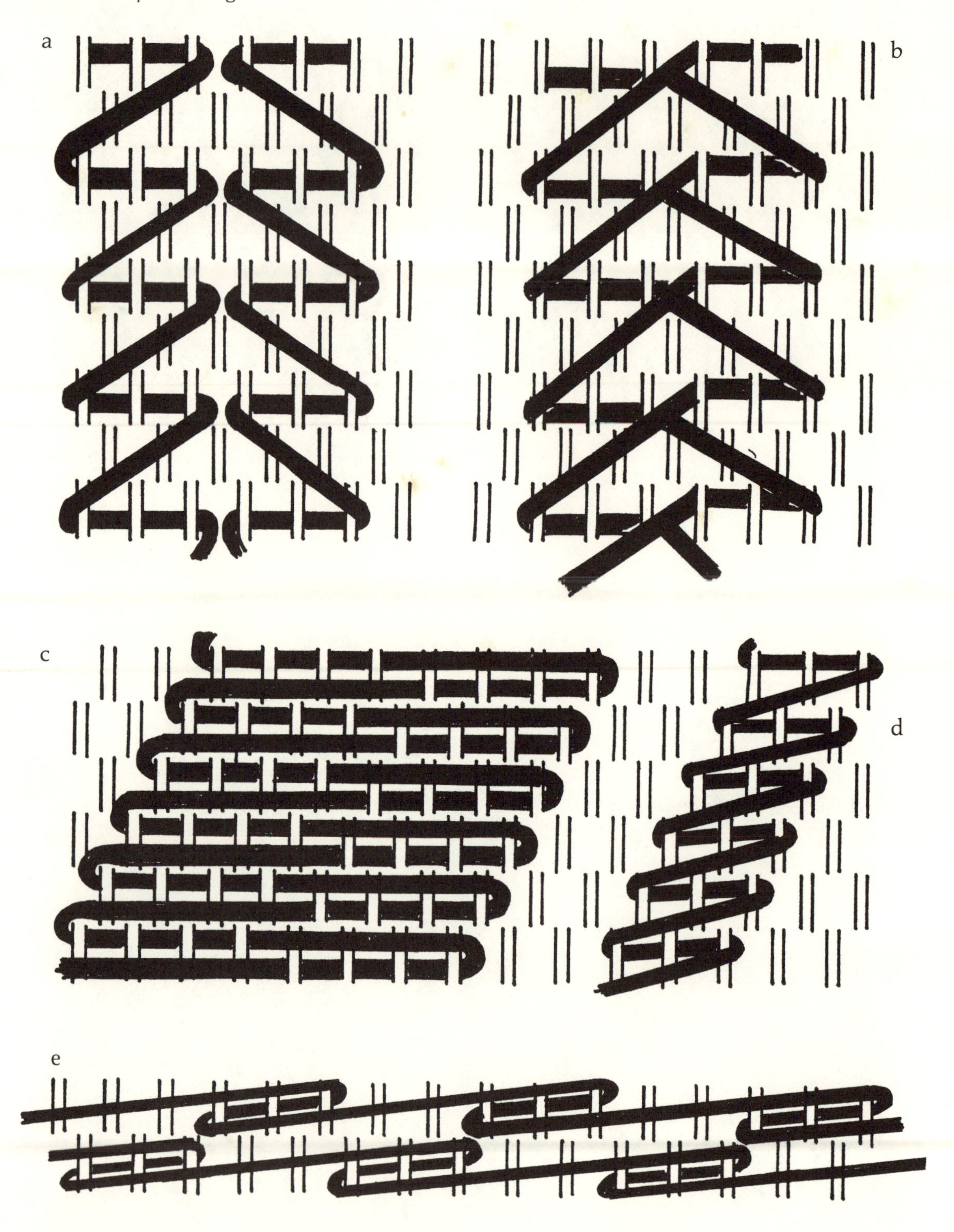

Fig. 86. Wrapped weaves, Nebaj: *a* and *b,* weave producing vertical lines; *c* and *d,* weave producing diagonal lines; *e,* weave producing horizontal lines.

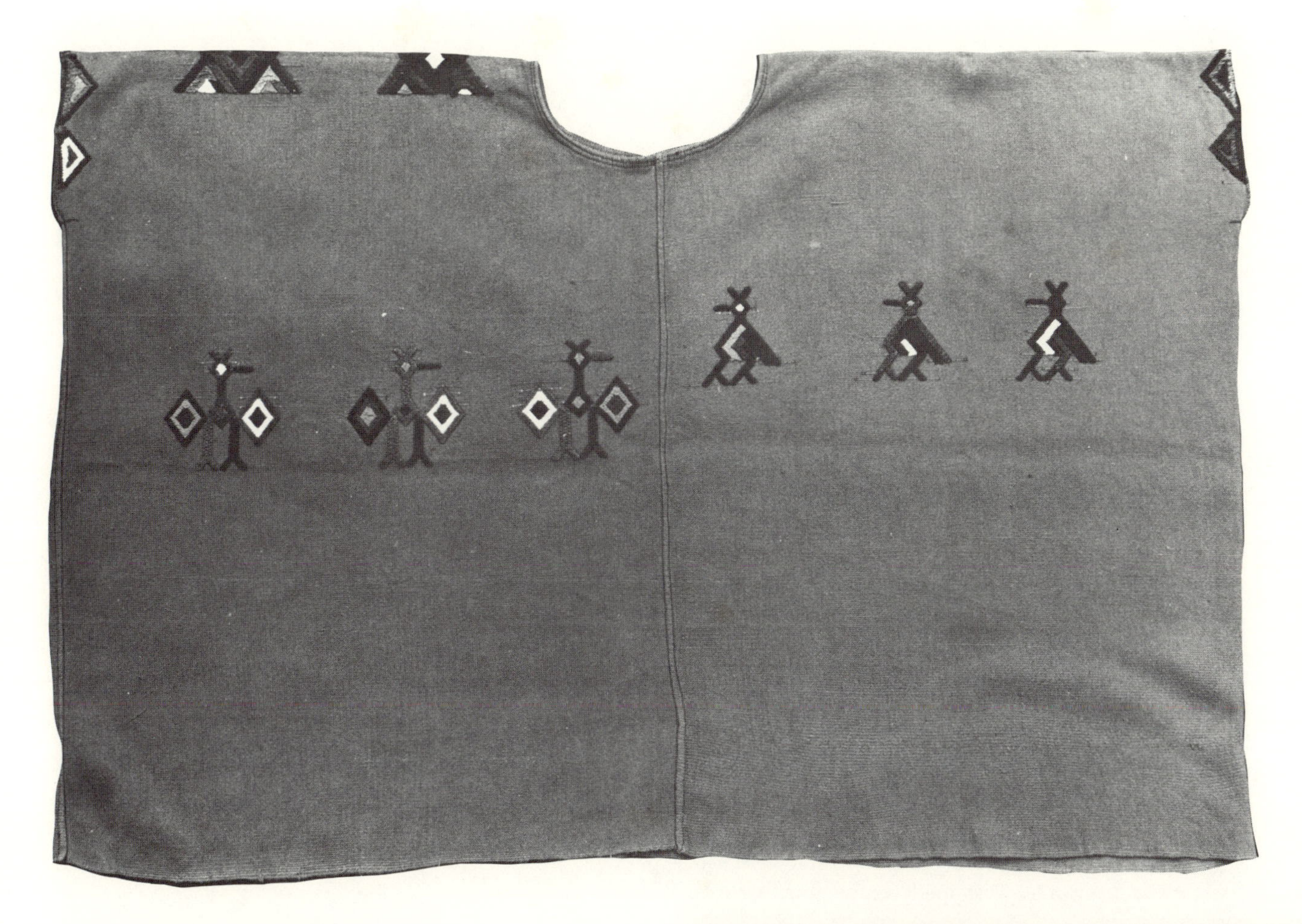

Plates 75 and 76. Huipil from Chajul.

acteristic of Nebaj weaving. In figure 86, warps and wefts are loosely spaced; in the actual weaving, they are close together. Loose ends of the extra wefts are put into the basic sheds, where they form weft cords. These weft cords, or streamers, can be seen in plate 74.

HUIPIL FROM CHAJUL

Special features: Single-face wrapped weave, single-face brocading (plates 75, 76).

Huipils in Chajul are carefully woven and the fabric, as a result of heavy battening, is very firm. The ground material is either red or white and the decorations consist of bird motives and geometrical forms.

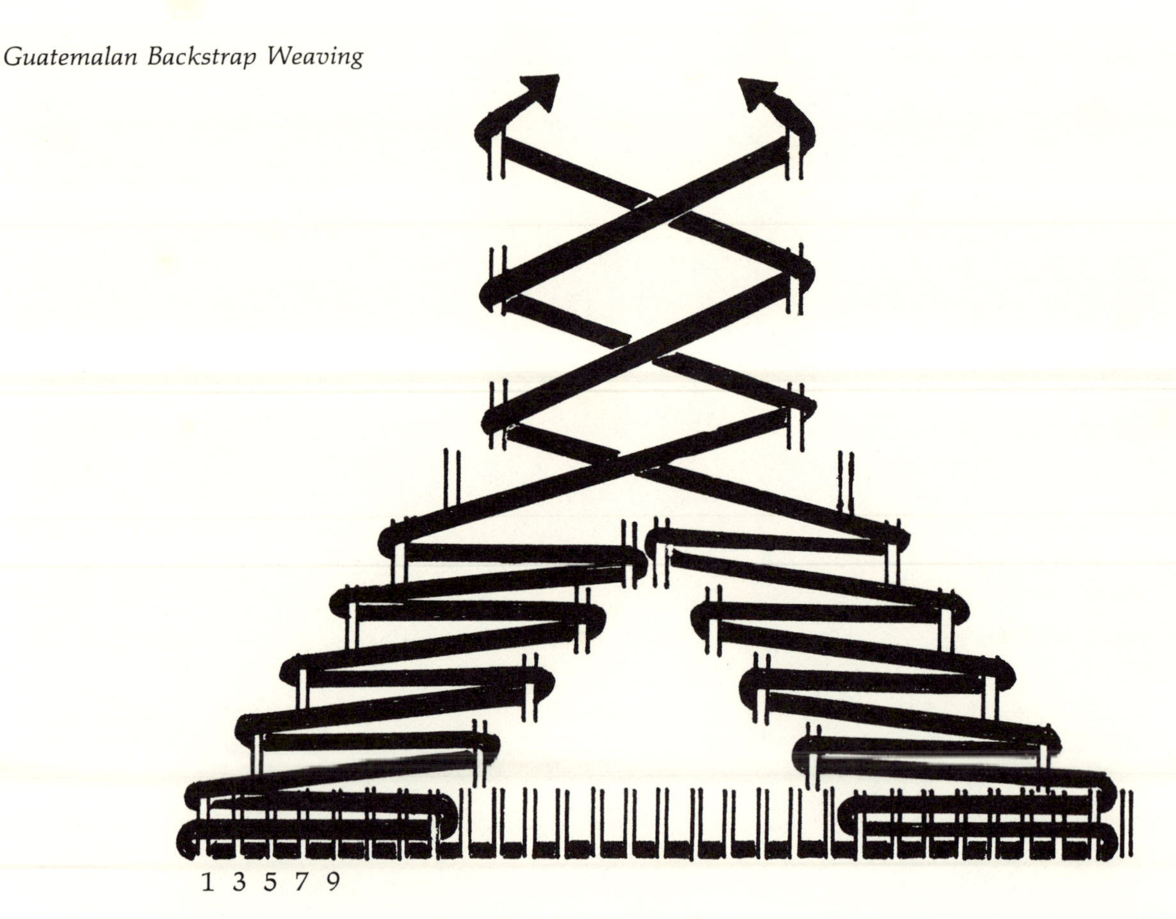

Fig. 87. Wrapped weave (bird's foot), Chajul.

The huipil shown in plate 75 is made from two panels that are joined with machine stitching. A round opening is cut for the head and hemmed in. The huipil is 22 inches long and 34 inches wide. Each panel has four selvages and a join. The warp consists of three red singles, the weft of four red singles. The thread count is 82 warps and 22 wefts per inch. The extra-weft patterns are done with two-ply yarn and in some sections with silk floss. The huipil is decorated with two rows of birds (one row on the back, one in the front). Along the shoulder fold is a row of geometrical motives. The arm holes are bordered by triangles. Each motif is composed of several sections of different colors, much like the figures on the Nebaj huipil described above. The weaving technique, however, is somewhat different from the one used in Nebaj. Figure 87 shows the foot of a bird (same as in plate 76) in detail. The two ends of each extra weft travel in different directions to build up the pattern. Compared to Nebaj patterns, fewer warps are employed to hold the extra wefts in place.

No special tools are needed for the wrapped weave. The ends of the extra wefts are put into the basic sheds, forming little weft cords (plate 76).

In plate 45, a girl from Chajul is shown weaving a row of birds. The pattern sticks in the loom are needed for weaving the horizontal bands.

WOMAN'S BELT FROM NEBAJ

Special features: Single-face wrapped weave, single-face brocading, warp stripes, tapered ends, and end fringes (plates 77, 78).

Weavers in Nebaj spend much time and effort weaving their colorful belts. The belt shown in

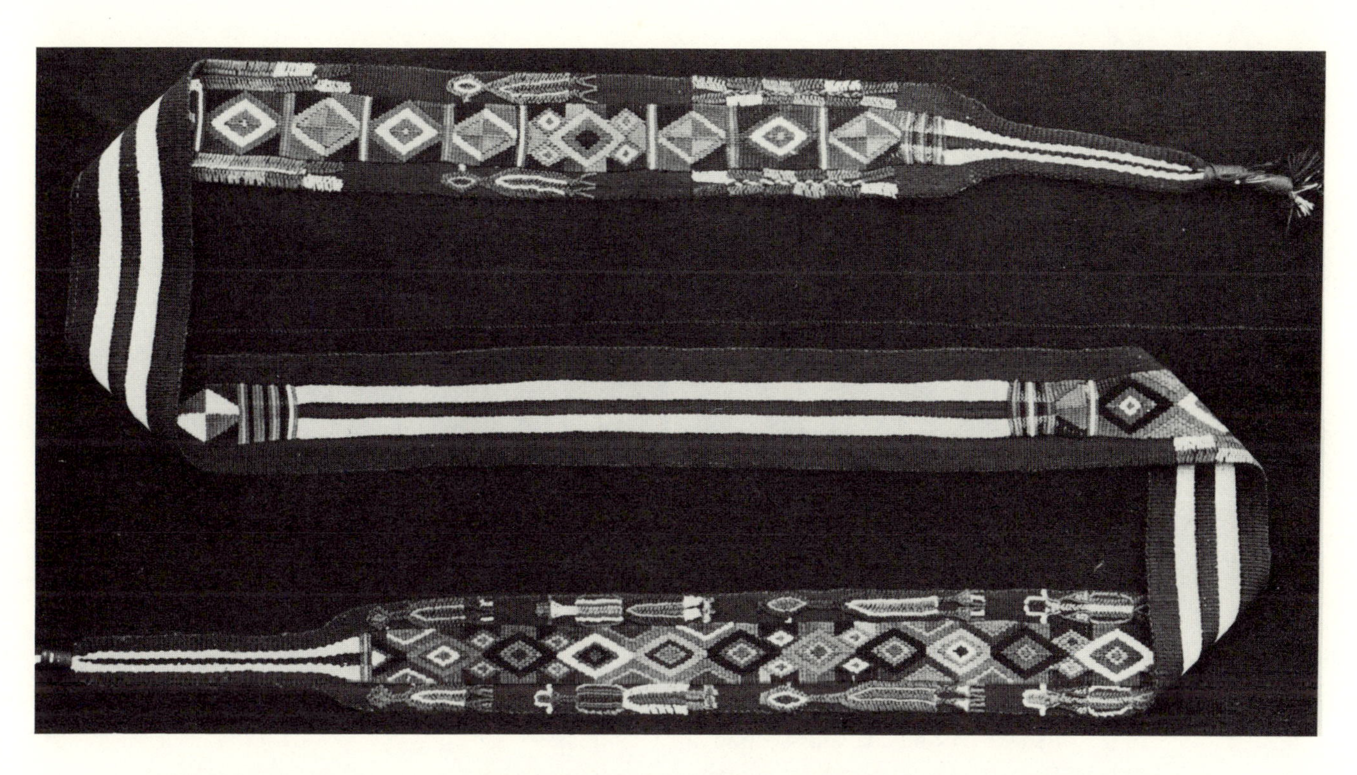

Plates 77 and 78. Belt from Nebaj.

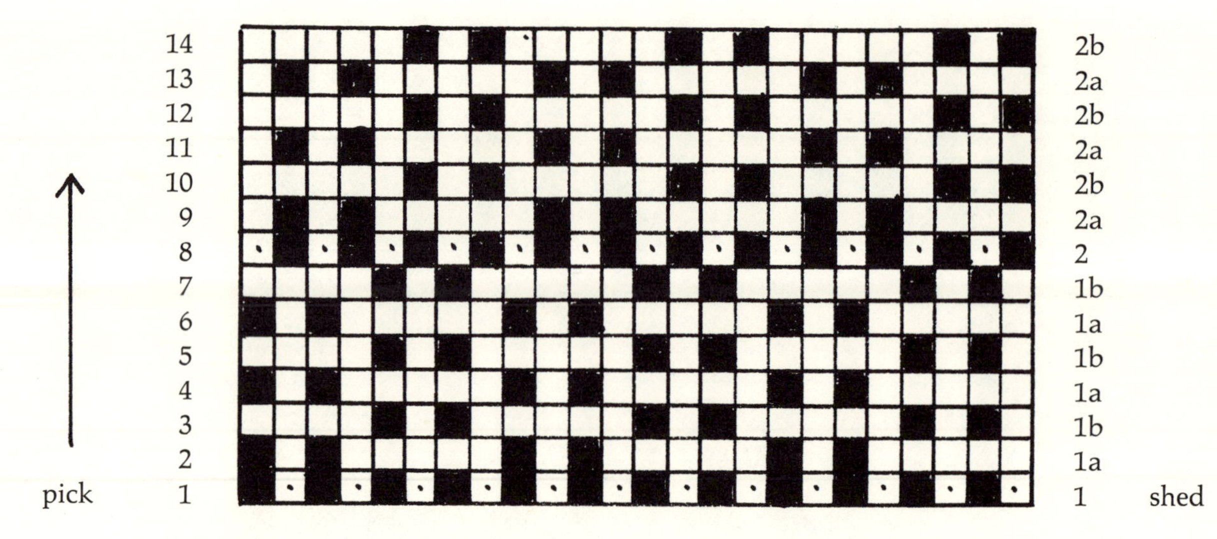

Fig. 88. Weaving sequence for extra-weft pattern. Black squares represent warps passing over a weft. White squares with a dot represent the basic weft. White squares represent the extra weft.

A basic weft in shed 1 is followed by six picks of extra weft, alternating in sheds 1*a* and 1*b*. Then come a pick of basic weft in shed 2 and six picks of extra weft in sheds 2*a* and 2*b*, and so on.

plates 77 and 78 is 3 yards long and 3 inches to 3⅜ inches wide. The ends are tapered and the last 3 inches of each end are left unwoven. At one end the warps have been cut. At both ends strings are wrapped around the bare warps to hold them together.

The warps are very tightly packed together, about 80 to 100 warps per inch. They are arranged to form two white bands and three red bands. The red bands are bordered by dark-blue stripes. The warps are arranged in the following order: 8 dark blue, 80 red, 8 dark blue, 36 white, 8 dark blue, 20 red, 8 dark blue, 36 white, 8 dark blue, 80 red, 8 dark blue—altogether 300 warps. The red warps consist of four singles twisted together, the white and blue warps are three-ply yarn. For the weft twelve red singles are combined. This makes a very strong weft, and the weft count is only 10 wefts per inch, even though the wefts are heavily battened down. For the extra wefts, two-ply yarn and mercerized cotton are used. Since the extra wefts need additional space where they pass between

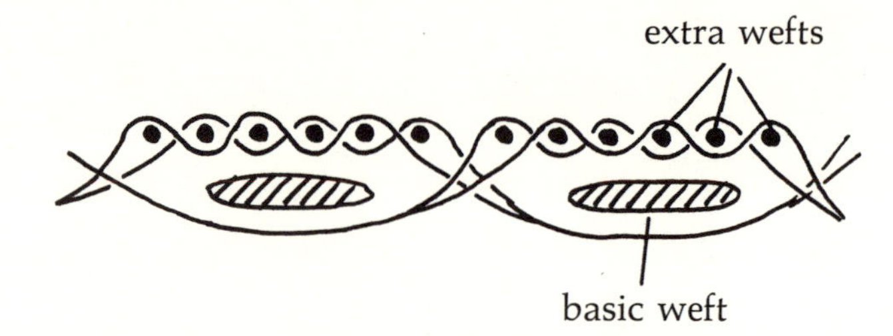

Fig. 89. Extra-weft pattern, cross section cut parallel to the warps.

warps, the belt is wider in sections that are decorated with extra wefts.

To weave the tapered ends of the belt, eight consecutive odd- (or even-) numbered warps were bundled together and treated as one warp. We did not see that part of the weaving, but assume that a special heddle was used.

Plate 79. Weaving a belt in Nebaj.

Extra-Weft Patterns

The outer red bands of the belt are decorated with birds, human figures, and corn plants in single-face wrapped weave. Two such birds can be seen in plate 78. Since the red bands are less than an inch wide, the figures are elongated and hard to recognize. The wrapping technique used for weaving the figures is much like the procedure described for the two-panel huipil from Nebaj.

The inner three bands of the belt are covered with geometrical patterns—diamonds, chevrons, horizontal bands—with a tapestrylike weft surface. To create an extra-weft surface, the weaver employs a special procedure. Each basic weft is followed by six or more extra wefts (or rows of extra wefts). Heavy battening pushes the extra wefts close together, so that they cover the warps.

The block diagram in figure 88 indicates the sequence in which the different sheds are established. It is assumed that the weaver wants to make a horizontal band, such as the one that separates the diamond shapes in plate 78.

Figure 89 shows a cross section of the material cut parallel with the warps to illustrate the position of the extra wefts on top of the basic weft. Only cross sections of the wefts can be seen.

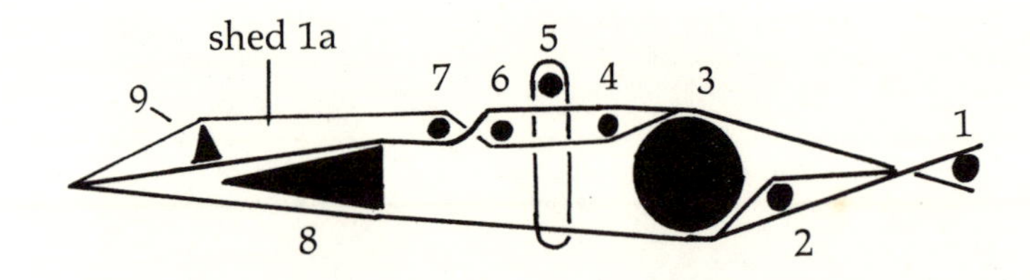

Fig. 90a. Weaving of extra-weft pattern, shed 1*a*.

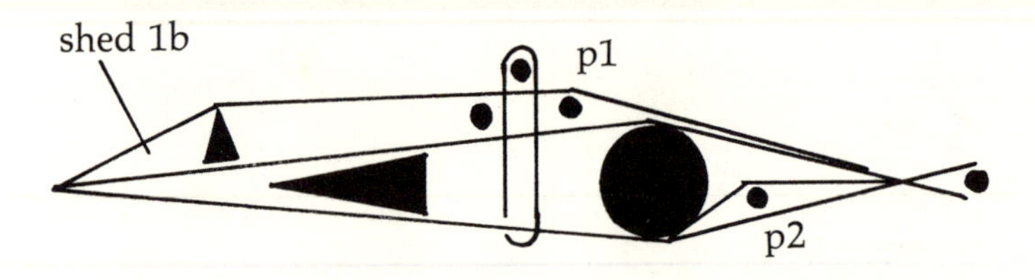

Fig. 90b. Weaving of extra-weft pattern, shed 1*b*.

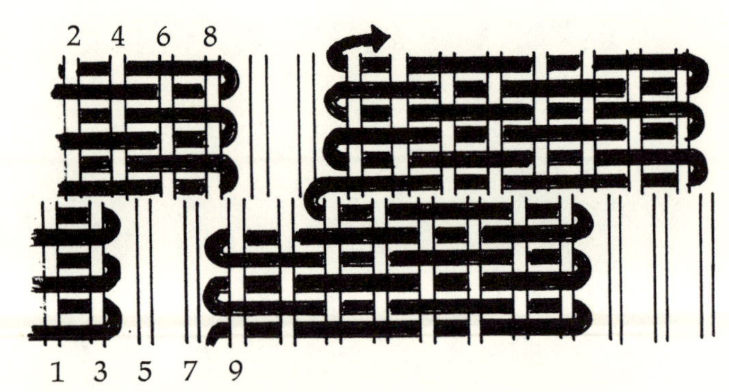

Fig. 91. Section of a diamond pattern, extra wefts of different colors separated by two warps. Basic wefts omitted.

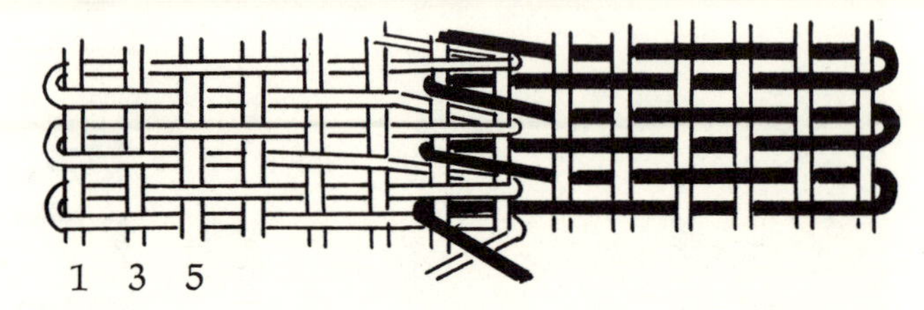

Fig. 92. Section of a diamond pattern, overlapping extra wefts.

Pattern Sticks

Sheds 1*b* and 2*b* are established with the aid of pattern sticks *p*1 and *p*2. These sticks are permanently kept in the loom. Sheds 1*a* and 2*a* are established with sticks that are temporarily inserted into the warps.

Weaving a belt is shown in plate 79. Starting at the top of the loom, the following loom sticks can be distinguished:

1. Lease rod with hooked end.
2. Pattern stick *p*2, dividing the even-numbered warps. The red warps at the edges of the belt all pass under *p*2.
3. Shed roll of rather large diameter.
4. Pattern stick *p*1, positioned close to the heddle. This stick divides the odd-numbered warps (two over, two under), except for the red warps at the sides, which all pass under *p*1.
5. Heddle with a hooked heddle stick.
6. Pattern stick *p*1.1, used to transfer the shed that is made by *p*1 (shed 1*b*) to the front of the heddle.
7. Pattern stick *p*1.2, used to establish shed 1*a* (countershed to 1*b*).

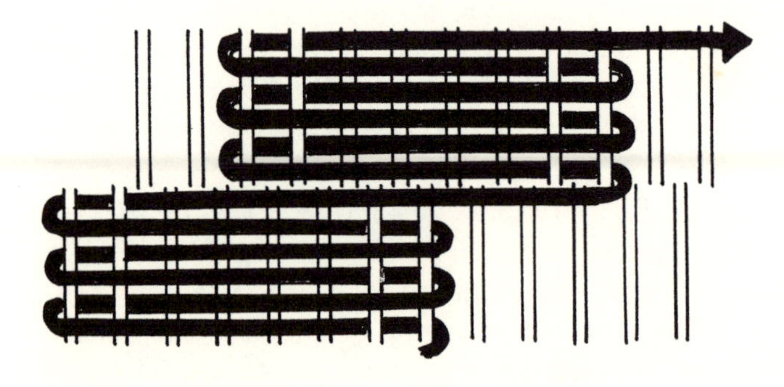

Fig. 93. Detail of a diamond pattern, Nebaj.

8. Batten, positioned in shed 1.

9. Small brocading sword, set on edge. The brocading sword holds open the shed that is established by *p*1.2.

10, 11. Lower end bar and cloth bar. The warps are tied to the end bar.

The same placement of the loom sticks is also shown in figure 90*a*.

Weaving Sequence

Basic Weft in Shed 1

The weaver puts a basic weft in shed 1 and battens it down. The batten remains in shed 1.

Extra Wefts in Sheds 1a and 1b

Pattern stick *p*1 is moved close to the heddle to make shed 1*b*. In front of the heddle, the weaver inserts stick *p*1.1 into the shed that is made by *p*1. In order to make shed 1*a*, the weaver uses a brocading sword to pick up the odd-numbered warps that pass *under* *p*1. She places stick *p*1.2 under these warps and positions it close to stick *p*1.1. The brocading sword is moved close to the working edge of the weaving and set on edge (see plate 79 and figure 90*a*). The weaver then puts an extra weft into shed 1*a*. She pushes the extra weft towards the edge of the weaving with the brocading sword.

To establish shed 1*b* for the next pick of extra weft, she removes the brocading sword and stick *p*1.2 from the loom. The brocading sword is reinserted under the warps that pass over *p*1 and set on edge, thereby opening shed 1*b*. The extra weft is now placed in shed 1*b*. Sheds 1*a* and 1*b* are used in turns for the extra weft. Altogether, the weaver makes about six picks of extra weft before putting the next basic weft into shed 2.

Basic Weft in Shed 2

The weaver pushes *p*1 behind the shed roll and removes *p*1.1, the batten and the brocading sword from the loom. She establishes shed 2, battens down the warp cross and puts a basic weft in shed 2. The batten remains in shed 2.

Extra Weft in Sheds 2a and 2b

The weaver moves pattern stick *p*2 close to the heddle and transfers shed 2*b* to the front of the heddle, where stick *p*2.1 is inserted into the shed. Sheds 2*a* and 2*b* for the extra weft are then established in the same way as described for sheds 1*a* and 1*b*. However, the weaver now works with the even-numbered warps.

So far we have described the weaving of horizontal bands. Other patterns—diamond or chevron shapes—are built up from little rectangles of different colors. Different sections of a shed are occupied by extra wefts of different colors. Otherwise, the weaving sequence remains the same as described above.

Figure 91 shows a small section of a diamond pattern. There are two bare warps between adjacent rectangles of different colors. Because of this arrangement, the rectangles are clearly separated from each other. In plate 78 the white warps can be seen in the gaps between the rectangles.

In the same illustration we notice that within one of the diamond shapes extra wefts of two different colors seem to overlap. A section of this pattern is shown in figure 92.

Loose ends of the extra wefts are hidden in the basic sheds.

Weavers in Nebaj decorate their belts with diamond and chevron patterns, often building these patterns from small rectangular elements. Figure 93 shows a detail of a pattern from another belt. Compared to the pattern shown in figure 91, the extra wefts form longer floats. Other than that, the resulting pattern looks very similar.

WOMAN'S BELT FROM SANTIAGO CHIMALTENANGO

Special features: Warp stripes, warp floats, and braided end fringes (plates 80, 81, 82).

The belt shown in plate 80 is 3 yards, 10 inches long, including the braided end fringes, and 4⅛ inches wide. It is decorated with simple warp-float

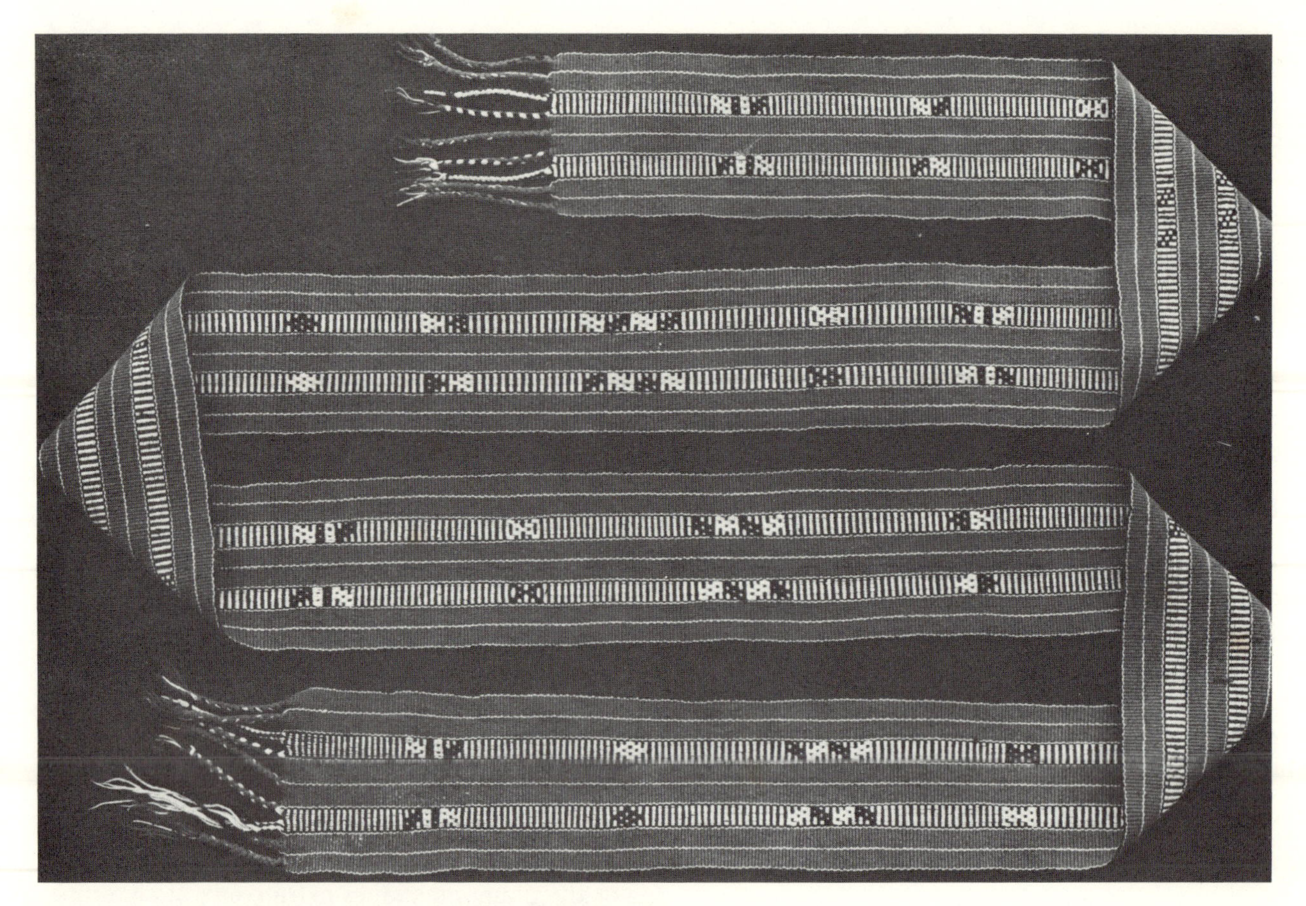

Plates 80 and 81. Belt from Santiago Chimaltenango.

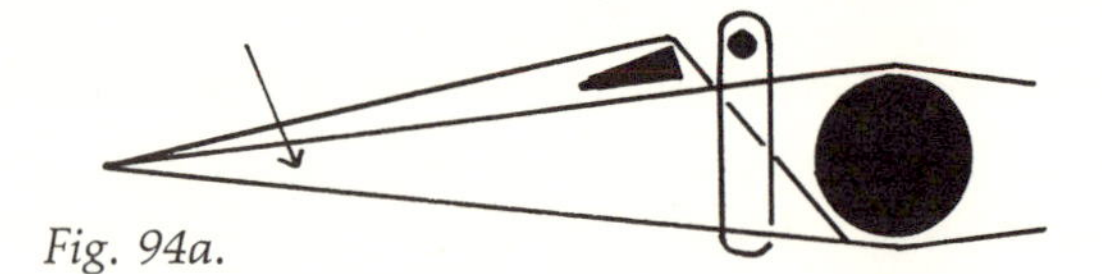

Fig. 94a.

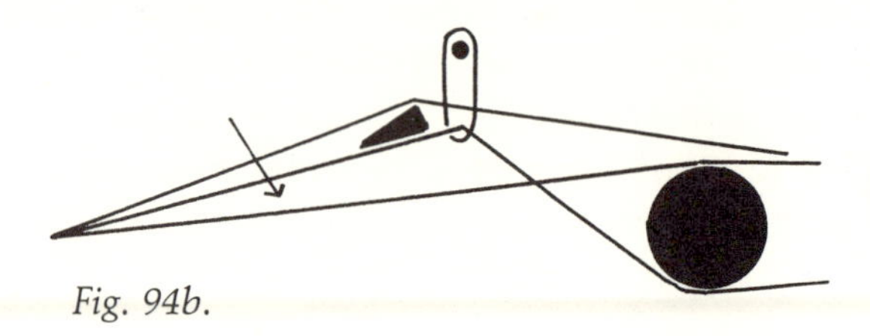

Fig. 94b.

patterns and warp stripes. Extra wefts are not used. There are three kinds of warps:

1. Hand-spun, dark brown wool, two strands twisted together.
2. Five red cotton singles, twisted together.
3. Four white cotton singles, twisted together.

The weft consists of two strands of dark-brown wool twisted together. The thread count is 60 warps per inch (for the red warps) and 12 wefts per inch.

The belt has two bands that are done with odd- and even-numbered warps of different colors. In each band, ten white and ten dark-brown warps alternate, forming cross stripes. About every four inches warp float patterns occur within these two stripes. If all the warps were of the same color, such warp floats would not stand out against the background. Therefore, weavers use two sets of warps of contrasting colors. Two of the warp float patterns appear in plate 81. All the warp floats are of equal length, passing over three wefts. To add more body to the pattern, the weaver always picks up two warps together to form warp floats.

Weaving Procedure for the Warp-Float Pattern

Let us assume that the dark warps are even-numbered (heddle-controlled) and the white warps are odd-numbered. The last step in plain weaving was putting a weft in shed 2. The batten is still in shed 2. Now the weaver wants to start the warp floats.

Step 1

With a brocading sword, she picks up the dark warps that are to form warp floats. She moves the brocading sword close to the heddle and withdraws the batten from the loom (figure 94*a*).

Step 2

She inserts the batten under all odd-numbered warps and under the even-numbered warps that pass over the brocading sword. The arrow in figure 94*a* indicates the shed into which the batten is placed. The weaver battens down the warp cross, sets the batten on edge, and puts a weft into the shed. The weft is battened down.

Plate 82. Couple from Santiago Chimaltenango.

Plates 83 and 84. Belt in Totonicapán style.

Step 3

The weaver withdraws the brocading sword from the loom and inserts it under the odd-numbered warps that are to form warp floats. The batten is withdrawn.

Step 4

The weaver lifts up the heddle, as shown in figure 94*b*. She inserts the batten under the even-numbered warps and the odd-numbered warps that pass over the brocading sword. The arrow in figure 94*b* indicates the shed where the batten is inserted. The warp cross is battened down, the batten is set on edge, and a weft is placed into the shed. The weaver battens down the weft and withdraws the brocading sword. Next comes step 1.

WOMAN'S BELT IN TOTONICAPAN STYLE

Special features: Warp-float patterns combined with extra wefts, braided ends (plates 83, 84).

Belts from Totonicapán are worn in many towns and copied by weavers in different places. They are decorated their entire length with intricate warp float patterns. The belt in plate 83 shows typical Totonicapán patterns such as hands, birds and human figures, and in between, the word "GLASS." The patterns consist of black and white warp floats and extra wefts of different colors that are inserted under the warp floats.

The belt is 2 yards, 14 inches long, excluding the braided ends. The ends are badly damaged. The width is 3⅜ inches. Three kinds of warps are used:

1. Red three-ply yarn for the outer edges of the belt.
2. Two white singles twisted together.
3. Two black wool singles twisted together.

The even-numbered warps are black, the odd-numbered warps white (except for the red warps at the edges). For the weft, four white singles are twisted together. The thread count is 100 warps and 23 wefts per inch.

Silk floss of different colors was used for the extra wefts. Most of the extra wefts have disintegrated, because the belt is old and worn.

We find two kinds of warp-float patterns on the belt. In the first, used to make geometrical shapes, only the black warps form floats. The second kind of pattern is produced by floats of the black and white warps. In both the floated warps always come in pairs. Each float passes over three basic wefts. The warp floats also hold down extra wefts.

We observed Totonicapán-style belts being woven in the town of San Antonio Aguas Calientes, where weavers have learned to copy textiles from many places. The belts look exactly like the ones from Totonicapán, but it is possible that Totonicapán weavers use a different loom setup. Lilly de Jongh Osborne, in her book *Indian Crafts of Guatemala and El Salvador*, shows a belt loom from Totonicapán with three heddles. Weavers from San Antonio do *not* use extra heddles for weaving the belts.

Weaving Procedures

Since belts have end fringes, it is not necessary to tie the warps to the end bars. The first few weft picks are strongly battened down, so that they are pushed close to the lower end bar. The weaver then leaves a few inches of warp unwoven—these warps will be braided at the end—and begins weaving the main part of the belt. After two inches of plain weaving, she makes the first pattern. On the belt shown in plate 83, the zigzag pattern consists of extra wefts that are held in place by floats of the black warps. The weaving sequence for this kind of pattern is as follows:

Weaving Sequence: Pattern with Extra Wefts Held Down by Floats of the Even-Numbered Warps

The pattern is started after a basic weft has been put in shed 2. The weaver battens down the weft

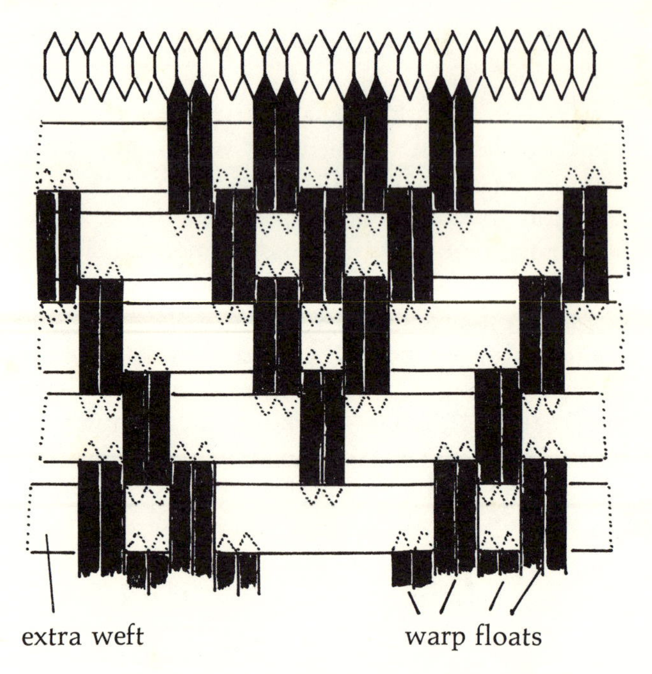

Fig. 95. Warp floats and extra wefts on a belt in Totonicapán style.

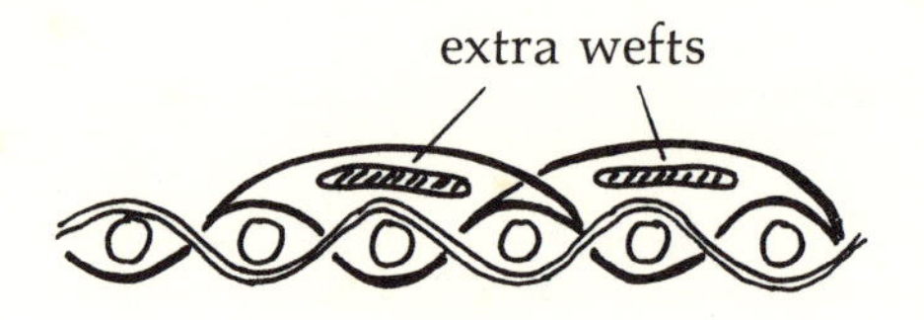

Fig. 96. Warp-float pattern, cross section.

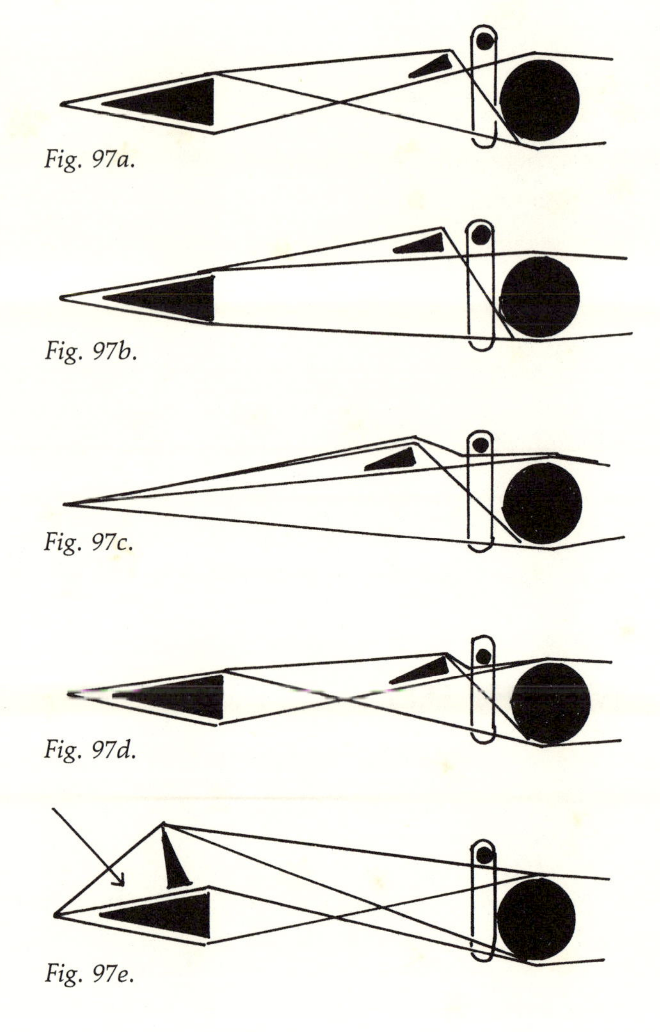

and leaves the batten in shed 2. All the black warps pass over the batten.

Step 1

The weaver inserts the brocading sword under pairs of black warps that are needed for the warp floats. She pushes the brocading sword close to the heddle.

Step 2

The batten is withdrawn from shed 2 and inserted under all odd-numbered warps and the even-numbered warps that pass over the brocading sword. The batten is set on edge to make a shed for the basic weft. The weaver puts a basic weft into the shed and battens it down. The batten is then taken out of the loom. The brocading sword remains in place.

Step 3

The weaver lifts the heddle to raise the even-numbered warps. She inserts the batten in shed 2, sets it on edge and makes a pick of the basic weft. She battens down the basic weft and leaves the batten in shed 2.

Fig. 98. Warp floats of black and white warps on a belt in Totonicapán style.

Step 4

The brocading sword, which in step 1 was inserted under some of the even-numbered warps, is now moved close to the working edge of the weaving and set on edge. The weaver puts an *extra weft* into the shed and battens it down so that it lies on top of the white (odd-numbered) warps. The brocading sword is then taken out of the loom, and the weaver returns to step 1.

Figure 95 shows a section of the zigzag pattern that can also be seen in plates 83 and 84. In plate 83, it is the first pattern at the left end of the belt.

Figure 96 shows a cross section of the material cut parallel to the warps. The cross sections of the *basic* wefts appear as circles. The cross section of the *extra* wefts is more elongated.

The second pattern on the belt—a row of five men wearing hats—is built up by floats of the *black and white* warps and by extra wefts.

Weaving Sequence: Second Pattern

The pattern starts at a point where the batten is in shed 2, and all the black warps pass over the batten.

Step 1

The weaver inserts the brocading sword under the pairs of black warps that are needed for the warp floats. She pushes the brocading sword close to the heddle (figure 97*a*).

Step 2

The batten is withdrawn from shed 2 and inserted under all the white (odd-numbered) warps and under the black warps that pass over the brocading sword (figure 97*b*).

The weaver sets the batten on edge and puts a basic weft into the shed. The weft is battened down.

Step 3

Using a bone needle the weaver picks up pairs of white warps that are needed for warp floats. The bone needle also picks up the black warps that pass over the brocading sword. The brocading sword is then inserted alongside the bone needle, and the needle is taken out of the loom. The batten is also taken out (figure 97*c*).

Step 4

The brocading sword is still positioned under the black and white warps that will form the warp floats. The weaver lifts the heddle, thereby raising all the black warps and the white warps that pass over the brocading sword. She inserts the batten into the shed (figure 97*d*).

The warp cross is battened down, the batten is set on edge and the weaver makes a pick of the basic weft. She battens down the weft and leaves the batten in the shed.

Step 5

The brocading sword is moved close to the working edge of the weaving and set on edge (figure 97*e*).

The weaver puts an extra weft into the shed and battens it down with the brocading sword. She removes the brocading sword but leaves the batten in shed 2. Next comes step 1.

Figure 98 shows a small section (man's head and hat) of the pattern that also appears in plate 84. Compared to the pattern shown in figure 95, the floats of the extra wefts are shorter. In the actual weaving they appear as dots between the warp floats.

Twenty-eight patterns on the belt are woven with warp floats of black and white warps. Eight sections are made with only black warp floats.

HUIPIL WOVEN IN CAJOLA STYLE FROM CONCEPCION CHIQUIRICHAPA

Special features: Twill weaving, combined single-face and two-face brocading, weft stripes (plates 85, 86).

Cajolá is a town near Quezaltenango, where most women have given up weaving but still wear a distinctive costume. The material for their huipils is woven either on treadle looms or by backstrap weavers in the nearby town of Concepción Chiquirichapa. The huipil shown in plate 85 was woven in Concepción Chiquirichapa and offered for sale on the market. It is 36 inches long and 39 inches wide. Two panels are sewn together and a hole is cut for the head. The huipil is left open at the sides. One of the panels has three selvages, the other has two. The cut edges are not hemmed. It is up to the buyer of the huipil to give it a better finish.

Three red singles are combined for warps and wefts. For the weft stripes, the twill weaving, and the extra-weft patterns, two-ply yarn in blue, green, yellow, and white is used. The thread count is 34 warps and 24 wefts per inch, and the fabric—because of the low warp count—is not very firm. Like so many items that are made for sale, the huipil is rather sloppily woven.

The front and back of the huipil have the same designs. At the lower end there are six weft stripes in blue, green, and yellow. On the upper half of the huipil, twill stripes alternate with brocaded patterns. To weave the twill stripes, the loom was equipped with a twin heddle, a pattern stick, and an extra heddle. The weaving procedure has been explained in chapter 19, under "Looms Equipped for Twill Weaving." The pattern stick required for twill weaving is also employed for the brocading. It passes under every other odd-numbered warp. Every basic weft that is put in shed 1 is followed by a row of extra wefts. The extra wefts are placed into sections of the shed that is established by the pattern sticks. When the extra wefts reverse their direction (at the border of a pattern), they go to the underside of the material, as in two-face bro-

Plate 85. One panel of a huipil in Cajolá style woven in Concepción Chiquirichapa.

Plate 86. Section of a huipil in Cajolá style.

cading. Extra wefts from adjacent figures interlock on the underside of the material, as described in chapter 20, under "Huipil from Concepción Chiquirichapa." Knots are tied at the beginnings and ends of the extra wefts and they are also placed on the underside of the material.

HUIPIL FROM TODOS SANTOS CUCHUMATAN

Special features: Horizontal bands in compound weave, single-face weft patterns, warp stripes, appliqué (plates 87, 88).

Huipils from Todos Santos are decorated with horizontal bands that have a weft surface and resemble tapestry. On old-style huipils, like the one shown in plate 87, plain red bands alternate with rows of small geometrical forms. The latest fashion is to cover large parts of the huipil with geometric designs. In both instances, the weaver employs a technique that produces two layers of wefts. This

Plates 87 and 88. Huipil from Todos Santos Cuchumatán.

compound weave is also used in the nearby town of San Juan Atitán.

Huipils from Todos Santos consist of three panels, and each panel has four selvages. The huipil shown in plate 87 is 27 inches long and 35 inches wide at the bottom. The center panel is 13 inches wide, the side panels are 11 inches wide. The panels become narrower in the section that is woven in compound weave. The sides of the huipil are closed, except for the armholes. A piece of white material and a strip of rickrack are sewn around the opening for the head, forming a collar.

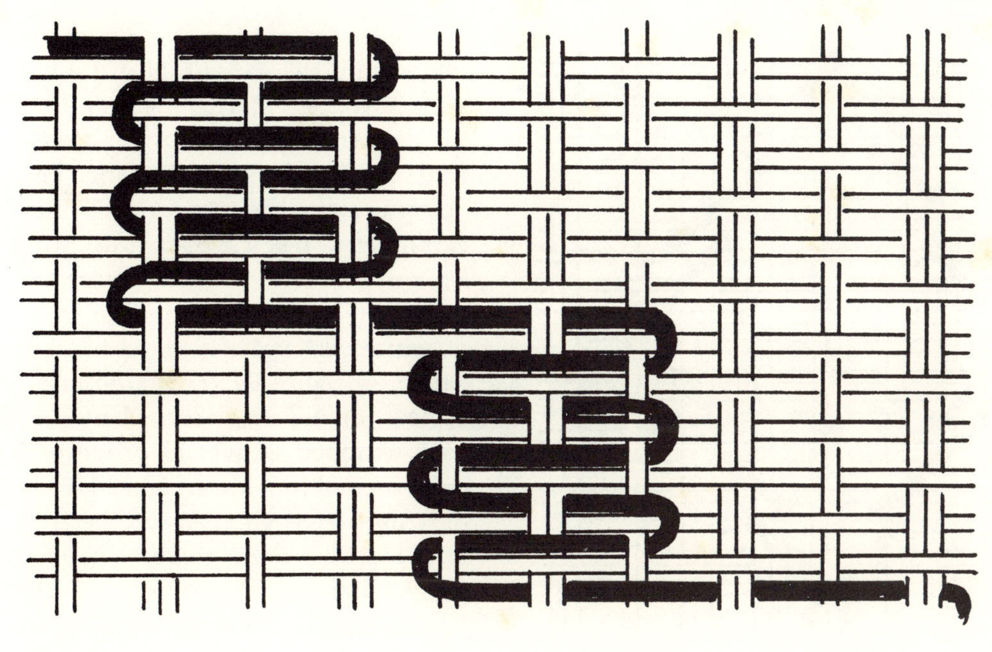

Fig. 99.

Warps

The white warps are singles that have been hand-twisted by the weaver to give them more strength. Hand-twisted yarn tends to form kinks, especially in the join area and at the end selvages. Red, orange, and blue stripes are woven with paired singles. The warp count is about 66 warps per inch for the white warps. Each panel is decorated with four wide and three narrow warp stripes. The wide stripes consist of the following warps: 3 red, 3 white, 3 red, 3 white, 25 red, 2 orange, 3 red, 3 blue, 3 red, 2 orange, 25 red, 3 white, 3 red, 3 white, 3 red. The narrow stripes are made up of 2 blue, 16 red, 2 blue warps. There are 56 white warps between the stripes of the center panel and 48 white warps between the stripes of the side panels.

Wefts

Plain weaving is done with paired white singles. Several kinds of yarn are used for the compound weave. The lower layer of wefts consists of white singles. In the top layer we find two singles of red *hilo Alemán* combined and two-ply blue yarn as background for the extra-weft patterns. The weft count is 21 wefts per inch for the plain weaving. The compound weave has a weft count of 84 wefts per inch for the upper layer and 28 wefts per inch for the lower layer.

Compound Weave

A detailed description of the weaving procedure is given in chapter 19 under "Loom for a Compound Weave." The sequence of weft picks is six or more wefts of the top layer, followed by two wefts of the lower layer. Much of the compound weave is done with *hilo Alemán*, an imported yarn that has disappeared from the market. It produces a velvety surface of vivid red color.

Extra-Weft Pattern

Superimposed over the compound weave are rows of geometrical patterns (see plate 88). These patterns are composed of little rectangles. Two-ply yarn and acrylic wool are used for the extra weft.

Fig. 100.

Wherever required by the design, extra wefts are placed into the sheds that are established for the surface wefts. Figure 99 shows a small section of the pattern. Only the surface wefts and the extra wefts are shown in the drawing. The ground wefts are omitted, since they do not interlock with the extra wefts. The warps that do not interlock with the surface wefts are also omitted.

The huipil under discussion has only one kind of extra-weft pattern with inlaid extra wefts as shown in figure 99. This kind of pattern is fairly typical for older textiles. Many weavers also employ wrapped weaves to produce figures that are composed of thin lines. Figure 100 gives an example of wrapped weaving. Again, the extra weft is superimposed over the surface wefts of the compound weave. Many more patterns can be found on Todos Santos textiles. The girl shown in the color plate of children from Todos Santos on page 63 is wearing a huipil with two kinds of patterns. Two rows of zigzags are woven in a wrapped weave, with extra wefts added to the surface wefts. Three bands with geometric figures (lozenges and parallelograms) are woven in tapestry fashion. The weaver used discontinuous surface wefts to build up the pattern.

»»»»»»»»»X«««««««

WORKS CONSULTED

Albers, Anni. *On Weaving.* Middletown, Conn.: Wesleyan University Press, 1965.

Becker-Donner, Etta. "Die Bedeutung der Weberei in Dorfgemeinschaften Guatemalas [The significance of weaving in Guatemalan village communities]." In *Verhandlungen des 38. Internationalen Amerikanistenkongresses* 2:439–48. Munich: Kommissionsverlag Klaus Renner, 1968.

Bird, Junius B. "Suggestions for the recording of data on spinning and weaving and the collecting of material." *Kroeber Anthropological Society Papers,* no. 22 (1960), pp. 1–9.

Delgado, Hilda S. "Guatemalan Indian Handweaving: Conservatism and Change in a Village Handicraft." In *Verhandlungen des 38. Internationalen Amerikanistenkongresses,* 2:449–57. Munich: Kommissionsverlag Klaus Renner, 1968.

Emery, Irene. *The Primary Structure of Fabrics.* Washington, D.C.: The Textile Museum, 1966.

Fergusson, Erna. *Guatemala.* New York and London: Alfred A. Knopf, 1937.

Gayton, A. H.: "Textiles and Costume." In *Handbook of Middle American Indians,* 6:138–57. Austin, Tex.: University of Texas Press, 1967.

Hahn-Hissink, K. *Volkskunst aus Guatemala* [Folk art of Guatemala]. Frankfurt: Städtisches Museum für Völkerkunde, 1971.

Martinez Pelaez, Severo. *La Patria del Criollo.* Guatemala: Editorial Universitaria, 1973.

O'Neale, Lila M. *Textiles of Highland Guatemala.* Washington, D.C.: The Carnegie Institution, 1945.

Ortiz Maldonado, J. P. "El traje en una comunidad indigena [Costume in an Indian community]." *Guatemala Indigena* 7 (1972): 163–226.

Osborne, Lilly de Jongh. *Indian Crafts of Guatemala and El Salvador.* Norman: University of Oklahoma Press, 1965.

Sowards, Elizabeth. *The Guatemalan Huipil.* Guide to an exhibition held at The Textile Museum. Washington, D.C.: The Textile Museum, 1973.

Start, Laura E. *The McDougall Collection of Indian Textiles from Guatemala and Mexico.* Occasional Papers on Technology, no. 2. Oxford: Pitt Rivers Museum, University of Oxford, 1948.

Wood, Josephine, and Lilly de Jongh Osborne. *Indian Costumes of Guatemala.* Graz, Austria: Akademische Druck- und Verlagsanstalt, 1966.

»»»»X««««

INDEX